"Intense burnout is ironically the goal of neoliberal biopolitics – this innovative book on cyberpunk explores the temporality between the promises and the failures letting people slowly die in the accelerating shadows..."

Geoffrey Whitehall, *Acadia University*

"Caroline Alphin's book is on the leading edge of international political theory. It aptly tells the story of how neoliberalism produces new forms of social, technological, and embodied existence. Alphin pushes the reader to ask difficult questions about the taken for granted ways in which neoliberalism perpetuates itself via mechanisms ranging from the fitbit to the biohacker. It is an impressive book, which should be read by anyone interested in understanding the politics of modern cityscapes."

Jessica Auchter, *Associate Professor of Political Science, University of Tennessee Chattanooga*

Neoliberalism and Cyberpunk Science Fiction

Caroline Alphin presents an original exploration of biopolitics by examining it through the lens of cyberpunk science fiction.

Comprised of five chapters, *Neoliberalism and Cyberpunk Science Fiction* is guided by four central themes: biopolitics, intensification, resilience, and accelerationism. The first chapters examine the political possibilities of cyberpunk as a genre of science fiction and introduce one kind of neoliberal subject, the self-monitoring cyborg. These are individuals who join fitness/health tracking devices and applications to their body to "self-cultivate." Here, Alphin presents concrete examples of how fitness trackers are a strategy of neoliberal governmentality under the guise of self-cultivation. Moving away from Foucault's biopolitics to themes of intensity and resilience, Alphin draws largely from William Gibson's *Neuromancer*, Neal Stephenson's *Snow Crash*, Richard K. Morgan's *Altered Carbon*, along with the film *Blade Runner* to problematize notions of neoliberal resilience. Alphin returns to biopolitics, intensity, and resilience, connecting these themes to accelerationism as she engages with biohacker discourses. Here she argues that a biohacker is, in part, an intensification of the self-monitoring cyborg and accelerationism is in the end another form of resilience.

Neoliberalism and Cyberpunk Science Fiction is an invaluable resource for those interested in security studies, political sociology, biopolitics, critical IR theory, political theory, cultural studies, and literary theory.

Caroline Alphin is an English Instructor at Radford University. Her research interests include biopolitics, science fiction, genre studies, feminist theory, and studies of neoliberalism.

Neoliberalism and Cyberpunk Science Fiction
Living on the Edge of Burnout

Caroline Alphin

NEW YORK AND LONDON

First published 2021
by Routledge
52 Vanderbilt Avenue, New York, NY 10017

and by Routledge
2 Park Square, Milton Park, Abingdon, Oxon, OX14 4RN

Routledge is an imprint of the Taylor & Francis Group, an informa business

© 2021 Taylor & Francis

The right of Caroline Alphin to be identified as author of this work has been asserted by her in accordance with sections 77 and 78 of the Copyright, Designs and Patents Act 1988.

All rights reserved. No part of this book may be reprinted or reproduced or utilised in any form or by any electronic, mechanical, or other means, now known or hereafter invented, including photocopying and recording, or in any information storage or retrieval system, without permission in writing from the publishers.

Trademark notice: Product or corporate names may be trademarks or registered trademarks, and are used only for identification and explanation without intent to infringe.

Library of Congress Cataloging-in-Publication Data
A catalog record has been requested for this book

ISBN: 978-0-367-49099-7 (hbk)
ISBN: 978-1-003-04450-5 (ebk)

Typeset in Times NR MT Pro
by KnowledgeWorks Global Ltd.

Contents

Acknowledgments — viii

Introduction: Living on the Edge of Burnout — 1

1 The Neoliberal Science Fictions of Cyberpunk — 23

2 Self-Monitoring as Instrumentalized Self-Cultivation — 41

3 Subtle State Killing as a Mode of Neoliberal Governmentality — 65

4 Cyberpunk Necroscapes and Necro-temporality in *Blade Runner* — 86

5 Reframing the Biohacker Within the Logic of Intensity — 107

Conclusion: Defamiliarizing Neoliberalism Through Cyberpunk Science Fiction — 129

Index — 134

Acknowledgments

The early stages of this book began in 2017 as I wrote my dissertation proposal. My colleagues in the ASPECT program at Virginia Tech were unwavering supporters of this project. To Michael Zarella, my husband, I am grateful that you pushed me to audit a class that made this all possible. Your love, patience, support, intelligence, and respect were all integral in helping me reach the end. Thank you to Mauro Caraccioli, Rebecca Hester, Rohan Kalyan, and Phil Olson for your comments and suggestions, without which this book would not have been possible. And, I would especially like to thank François Debrix, my mentor and friend, for his unwavering support. François gave invaluable feedback and guidance throughout the drafting process. This project benefited from the numerous conversations I had with a number of my colleagues while at Virginia Tech, including Shelby Ward, Leigh McKagen, Mary Ryan, Linea Cutter, Mario Khreiche, and Tim Filbert. In addition, thank you to the friends and colleagues who provided feedback at conferences as fellow panelists, discussants, and chairs, and through various discussions and conversations. In these instances, I received strong support and feedback from Tim Luke, Mike Shapiro, Siobhan McEvoy-Levy, Kyle Grayson, Nick Kiersey, and Cara Daggett. I would like to thank my editor Natalja Mortensen, and her editorial assistant Charlie Baker, for their patience and guidance through this process and for believing in my project.

An earlier version of Chapter 3 appeared in the co-edited volume *Necrogeopolitics: On Death and Death-Making in International Relations*, Caroline Alphin and François Debrix, eds. (New York: Routledge, 2019). My thanks go to Robert Sorsby and Claire Maloney, and to volume contributors, Alex Barder, Ben Meiches, Jessica Auchter, Gitte du Plessis, Francine Rossone de Paula, Mauro Caraccioli, Brent Steele, Stephen Michael Christian, Mike Shapiro, Sam Opondo, Ali Musleh, and Kennan Ferguson for helping to make this project possible.

And, finally, thank you to my parents and my brother. I love you.

Introduction
Living on the Edge of Burnout

Introduction

It is a paradox that killing certain populations is a positive condition of biopower.[1] Biopower's primary functions are to define life as an object of governmentality, subject stochastic events to calculability, and make a population live.[2] And yet, it is also the case that these functions are tightly coupled with modes of state killing. Neoliberal governmentality[3] has utilized alternative modes of killing that have supplemented more traditional forms of human destruction such as state racism, spectacular killing, and the suspension of the law. As a way to eliminate surplus bodies that fail to function in the production of value, neoliberal governmentality has developed subtle ways of killing its subjects that often function as letting them die. Sometimes these modes of "letting die" are slow and quiet. Responsibilizing the individual disassociates health outcomes from systemic/state support, and produces an increase in insecurity and inequality, and therefore makes it easier to blame endemic health outcomes, such as obesity, on individual choice and/or poor character. At other times, these modes of "letting die," combined with more overt forms of state killing, are sudden and violent (e.g., a combined penal state[4] and new military urbanism[5] understand all populations as potential threats, but work towards eliminating surplus and threatening bodies, often through police violence and incarceration). This is not to suggest that neoliberal subjects are always individuals; rather, I utilize "the possessive individual," "the entrepreneurial self," "human capital," and "the individual" to refer to how neoliberal governmentality knows its subjects of power, produces these subjects, and develops techniques and technologies for regulating, governing, and managing these subjects as part of a population.

Neoliberalism must work to foster a reality whereby humans are driven by competition rather than exchange. Thomas Lemke argues that neoliberalism is "a political project that endeavors to create a social reality that it suggests already exists...," and "not only the individual body, but also collective bodies and institutions (public administrations, universities, etc.), corporations and states have to be 'lean,' 'fit,' 'flexible' and 'autonomous.'"[6] It appears as though neoliberal governmentality governs without governing as it spreads its subjectivity "over the surface of things and bodies, arouses it [subjectivity], draws it out and bids it speak, implants it in reality... reflected in a myriad of discourses, the obstination of powers, and the interplay of knowledge and power."[7] Neoliberalism is not just a political project. It is a general condition of existence that includes an array of political projects and it is a narrative project that fosters a particular reality. The interplay between these neoliberal discourses and nodes of knowledge-power produces a particular type of thinking subject: one that thinks of itself in economic terms, as an entrepreneur of itself. Individuals practice, and thus nurture, neoliberal subjectivity when they, for example, monitor their physiological performance in order to enhance their human capital, when they embrace insecurity in order to live intensely, when they accept all time, including leisure, as labor/capital time, or when they calculate their worth based on their ability to compete.

Part of the goal of this book, then, is to make it more difficult for neoliberalism to foster a reality that it "suggests already exits,"[8] and to make strange the values, governmental rationalities, and modes of knowledge production that neoliberalism takes for granted as just the way things are. One way this study does this is through a simultaneous engagement with cyberpunk science fiction and neoliberalism. Cyberpunk,[9] a subgenre of science fiction, brings into focus the sometimes slow and quiet and at times sudden and violent deaths that emerge out of a neoliberal system that shifts responsibility and the management of risk from the state onto individuals. Killing is not always catastrophic and overt. Many people do not recognize deaths in their community as a "form of state killing unless an agent of the state—such as a police constable—literally kills them. The cruddy, cumulative, and corrosive aspects of life have spread..."[10] too deep into the everyday. State killing can take the form of allowing individuals to live in lethal conditions and blaming those individuals for their inability to escape those conditions. The possessive individualism that motivates the main characters in cyberpunk highlights the kind of neoliberal subjectivity that understands its integrations with technology, its movements and lack of movements through space, and its encounters with risk as a result of its choices rather than as the product of systemic racism, economic inequality, and neoliberal governmentality. This is of course not to suggest that cyberpunk is simply reflective of the realities of neoliberalism, as though it were no more than a mirror image of the real. Rather, as I argue throughout this book, cyberpunk is also a productive force behind the perpetuation of neoliberal governmentalities such as the logic of intensity, resilience, competition, risk, and individualism, etc.

Thinking about neoliberalism through cyberpunk can offer a "refashioning of force relations" that points to the way neoliberal governmentalities, which function partly through the economization of everyday life, produce and benefit from increased precarity within contemporary cities. This economization of everyday life, in part, refers to what Foucault describes as the "economization of the entire social field":[11] a process that makes the social field (e.g., familial relationships, marriage, crime) intelligible only through the logic of the market. In addition, the economization of everyday life goes beyond generalizing the logic of the market throughout the social field into nearly all modes of living, which makes a population, and its space and time, intelligible via competition and the market. Precarity, on the one hand, refers to normalized forms of insecurity that, in tandem with racist discourses, make it easier to kill some individuals, while, on the other hand, it refers to normalized forms of insecurity that lead to slow, quiet, and seemingly blameless deaths. Precarity is a condition that biopower needs in order to function. Thus, thinking about neoliberalism through cyberpunk can "generate ways to think 'the political'"[12] that remain aware of the threat of death and bodily harm that some bodies face in cityscapes.

The scholarship that defines cyberpunk and analyzes the urban spatial and temporal realities that have influenced this science fiction genre has functioned as an echo chamber that simply reverberates what Jameson, Harvey, Bruno, and Boyer (among others) see in the social/cultural/political shifts of late-capitalism. Their analysis of cyberpunk as a cultural product that is embedded within economic processes is dependent upon "the maintenance of an all-encompassing, hegemonic paradigm, which is the cultural logic of late-capitalism."[13] Cyberpunk, along with their diagnosis of space and time, serves as evidence that the cultural shifts and social problems of late-capitalism are real. That is, cyberpunk is merely a representation, and affirmation, of reality, or a mirror image of the real.

My book moves beyond thinking about literature, in this case cyberpunk, in terms of interpretation and representation. I offer an analysis of cyberpunk that is itself a way to think the political that remains sensitive to the general conditions of existence under neoliberalism. I work against thinking about the cyberpunk genre as a representation of late-capitalism and as a tool for diagnosing this condition, and instead, I juxtapose cyberpunk genre archetypes/narratives and neoliberal discourses in order to make strange the taken-for-grantedness of these discourses and the general condition of competition. Rather than understanding the spatial and temporal changes that have emerged out of the development of information technologies, neoliberalism, and globalization as markers of postmodern aesthetics, this project works counter to the notions that space is borderless, boundless, motionless, or directionless, and that we have lost a stable subject, since the spatial and temporal conditions of competition under neoliberalism are still very much bordered and highly regulated by neoliberal governmentalities. Furthermore, by applying some of the methodologies provided by Michael Shapiro to the cyberpunk genre, I offer moments of interference and discomfort with dominant modes of intelligibility within neoliberalism, thereby opening the possibility for new subjectivities, for resisting normative frames, and for a politics that is more sensitive to the deadly nature of neoliberalism and its governmentality. In this study, I will juxtapose cyberpunk novels and films with political and economic genres.[14] This juxtaposition of genres can make visible ways of critically examining "power-city relationships"[15] that are often ignored in postmodern aesthetic readings of urban space, literature, and film. Gilles Deleuze writes that "...philosophical theory...is a practice of concepts, and it must be judged in the light of the other practices with which it interferes."[16] Juxtaposing these genres leads to epistemological interference, which can disrupt the status quo of neoliberalism.

Biopolitics Under Neoliberalism

Recent scholarship on biopolitics argues that Foucault's analytic for knowing power can no longer fully capture the ways life/living is governed and made an object of power. For example, Elizabeth Povinelli argues that there is a more fundamental binary (life/nonlife) that underpins biopolitics and better addresses other forms of governance that, for Povinelli, biopolitics misses.[17] Further, she suggests that scholarship on biopolitics is too focused on spectacular forms of death and biological racism, and thus misses ordinary suffering. This project works towards offering a biopolitics that theorizes death in terms of ordinariness/taken-for-grantedness and it suggests that biopolitics is still a useful analytic within neoliberalism. In other words, Foucault's biopolitics can do more than theorize a genealogy of biological racism and genocide. We do not necessarily need a separate analytic like necropolitics or thanatopolitics to understand the techniques, operations, and forces of biopolitics (something that I develop at length in Chapter 3). If, as I suggested above, killing is a positive condition of biopolitics, then biopolitics can account for the deployment of weapons of maximum destruction and the mechanization of biological racism, in addition to the new forms of social and bodily existence that may be unique to neoliberalism, which include differentially experienced endemics, self-entrepreneurship, and responsibilized calculation of risk. I accept that massive-scale killing is not the modus operandi of the biopolitics of neoliberal spatio-temporality, and that the logic of neoliberalism does more than "reduce social values to one market value,"[18] as Povinelli suggests. But, rather than advocating for moving beyond biopolitics,[19] I suggest instead that neoliberal biopolitics can still be understood in terms of Foucault's analytic.

Thus, this study cautions against moving beyond biopolitics fully and embracing a concept of necropolitics or thanatopolitics instead. That scholarship on biopolitics has focused on spectacular and extra-juridical death, or as Karen Yusoff suggests, "the generalities of this specific mode of government,"[20] should not suggest that we have to move beyond biopolitics in order to make visible the taken-for-granted/ordinary suffering and slow deaths that many individuals face under neoliberalism, as though biopolitics is in and of itself an insufficient analytic to understand how life is made an object of power. How power functions under biopower is intensified within neoliberalism as a population of individuals more efficiently governs/regulates itself and, at the same time, as the degree to which governing entities are beholden to its population significantly decreases. Biopolitics' integration of economic calculation into how we define and control life, makes it an especially useful tool for understanding how some neoliberal governmentalities function, since it is through economic calculation that a population of individuals becomes intelligible. Governmentalities that are driven by the logic of competition produce, maximize, and benefit from the increased everyday economic, political, social, and bodily precarity of a population. When some populations fail to compete and thus become surplus bodies, the spatio-temporal nature of neoliberalism supports a governmental rationality that interferes only when it is necessary to securitize and ensure competition. And one way it can securitize and foster the logic of competition is to let these populations of surplus bodies die.

This book questions the characterization of biopolitics as constituted by "outright extinction," rather than as "perseverance, endurance, effort, and precarious survival."[21] That is, under the logic of competition, the ontology of the neoliberal subject is defined by perseverance, endurance, effort, and precarious survival since within neoliberalism, subjects must compete, and thus, persevere through paying to work, enduring years of work without benefits and security, showing effort as they maximize their human capital, and embracing flexible work as the epitome of the free individual. While making live may be in theory the goal of biopolitics, in practice, living always works in tandem with competing. Most forms of life are precarious because the space and time of neoliberalism are hostile to social and political ontologies that are not subject to the logic of competition. Neoliberal governmentality produces the conditions of competition and then exonerates itself from any responsibility for these conditions; thus, a population of individuals is more and more on its own, and at the same time, has no choice but to accept and often praise this neoliberal reality.

In this book, I am especially interested in how neoliberal governmentalities shape and produce space and time, along with certain subjectivities, in order to secure and maximize market competition. I use the terms "necro-temporality" and "necroscapes" to describe and problematize the kind of temporalities and spatialities that emerge out of a generalized mode of existence that is defined by competition with others.[22] Unlike liberal biopolitics, the norm of neoliberal biopolitics is competition. There is an even more intense focus on the individual as a unit of population: a population that can be governed without being governed, since, according to the logic of neoliberalism, being is defined by living intensely, accepting life risks, maximizing human capital, and competing in the market. Thus, this population of individuals regulates itself according to the logic of competition. I argue that a proliferation and intensification of knowledge/power nodes (e.g., the privatization of the penal system, the health industry, the fitness industry, etc.) further certain morals and character traits (e.g., the responsibilized self, the competitive individual as a good in itself, the market as just, taking on life risks as morally correct, the fact that individuals that fail in the market are lazy and parasitical, a person's value is tied to the maximization of their human capital).[23] And, in tandem with these modes of knowledge/power included above, the law, the logic of competition, and often a new military urbanism produce necroscapes and necro-temporalities. That is, risk, responsibility, and fitness among other effects, are driven to excess (they are intensified) in spaces under what Stephen Graham has argued amounts to a new military urbanism,[24] an "urbanized" domain where all individuals are potential threats to securitizing market competition. Graham further develops Foucault's boomerang effect in his analysis of what he calls a new military urbanism. According to Graham, the violence, surveillance, tracking, identification, and targeting military practices used in the colonies and in other strategies of war have crossed over into civilian application. He writes that "[f]undamental to the new military urbanism is the paradigmatic shift that renders cities' communal and private spaces, as well as their infrastructure—along with their civilian populations—a source of targets and threats…new doctrines of perpetual war are being used to treat all urban residents as perpetual targets whose benign nature, rather than being assumed, now needs to be continually demonstrated."[25] I suggest also that, along with a militarization of space, there is a marketization of space that is shaped by a logic of competition. Like Graham's militarization of space, a marketization of space understands populations, more specifically populations of individuals, as a source of targets and as threats to competition. Civilians must prove their competitive worth (securing a job, maximizing their human capital, taking on risk) in order to move and live freely.

Biopower, Necropower, and Cyberpunk

Achille Mbembe defines necropower as the modern state's ability to reduce human life to its bare or instrumental forms in order to materially destroy human bodies. Necropolitics is the subjugation of life to the power of death. Mbembe is largely concerned with the ways "weapons are deployed in the interest of maximum destruction of persons and the creation of *death-worlds*."[26] If we understand necroscapes in the context of Mbembe's necropolitics (his perspective on biopolitics), they are territories (often cities) designed to capitalize on the reduction of human life to its bare forms by destroying unwanted bodies, or letting them waste away. For Foucault, biopolitics emerged out of immense political, social, and economic changes, which included early industrial capitalism and a rapidly increasing population. His work on biopower examines how power and the state adapted to these political, social, and economic changes through the development of governing rationalities that were concerned with managing large populations and insuring the production of capital could function smoothly. At its core, biopower and the techniques, governmentalities, and discourses that emerge out of its biopolitics are directed at making a population live. Once life became the object of power, governmental rationalities emerged that were concerned with endemics, the health problems of living in particular environments, and overall life-expectancies of a population. With biopolitics, we also saw an economization of everyday life as economic rationality bled into the non-economic, as I indicated above. Mbembe questions whether Foucault's biopower can still account for the ways the state and sub-state forces kill populations. His answer is that biopower fails to account for the ways modern sovereignty, the political, states, and sub-states subjugate life to the power of death. That is, biopower does not account for the spectacular modes of state and sub-state killing implemented in the colonies, Mbembe's primary site of study. According to Mbembe, Foucault's analysis of power was limited to western urban spaces and western modes of warfare and state killing.

Yet, Foucault does address the co-productive relationship between the spaces of the colonizers and the spaces of the colonized. In *"Society Must Be Defended,"* Foucault identifies this process as a "boomerang effect,"[27] which highlights the ways colonial militarism, including hyper-modes of surveillance and intense forms of violence, reverberated back into western urban discourses and security practices. Foucault writes that "[a] whole series of colonial models was brought back to the West, and the result was that the West could practice something resembling colonization, or an internal colonization, on itself."[28] This suggests that it is not so much that Foucault was interested only in western imperialism independent of its colonial practices; rather, it shows that Foucault was, in part, concerned with the ways securitization functioned as a technique of biopower. That Foucault does not address some of the necropolitical practices Mbembe examines does not mean that biopower cannot account for these spectacular modes of state and sub-state killing. Foucault's biopower is not limited to a particular form of sovereign power. Furthermore, the need to kill is not a contradiction within biopower, but is, instead, a positive condition for its perpetuation. Thus, biopower and biopolitics are still good analytics for understanding the ways power, at the level of the state, the population, or the individual, kills and/or allows to die. We do not necessarily need a new or different term to account for killing and letting die as both modalities are already incorporated within biopower. The time and space of death-worlds, both spectacular and mundane, can just as easily thrive within the techniques, strategies, discourses, and modalities of biopower/biopolitics. Thus, Foucault's biopower or biopolitics may better account for how killing functions as letting die in neoliberalism since it can highlight the seemingly natural and "irresponsible" ways some populations die.

Necroscapes, then, are constituted by different modes of state killing, and thus, a word like "destruction" does not encompass all of the ways biopolitics operate within these death worlds. Through neoliberal discourses, urban design and planning, securitization, terror, state violence, surveillance, and the production, engineering, and construction of vulnerable bodies, necroscapes kill and let die certain bodies. Necroscapes, under neoliberal governmentality, let bodies die by restricting access to health care, food, water, and other resources through the production of a possessive individualism that sees lethal habits and living conditions as purely a choice, as well as through systemic inequality. Through these spaces, neoliberal governmentalities use subtle and overt ways of "killing" unwanted bodies.

Mbembe is not explicit about the nature of time in necroscapes. He writes about different temporalities in necroscapes that produce life as "a form of death-in-life." Borrowing from Giorgio Agamben's *Homo Sacer*[29] and Carl Schmitt's notion of the state of exception,[30] Mbembe understands the modern state's ability to deny political subjectivity, reduce life to bare life, and exclude bodies from the "normal state of law" as setting the conditions for necroscapes. He sees plantation temporality, apartheid temporality, and colonial temporality as manifestations of the state of exception and as precursors to necroscapes. His understanding of certain bodies in necroscapes is that they are a kind of living dead; these bodies are from birth subject to the state's power to let die or kill.

I understand necro-temporality as a form of time in neoliberalism. It is a time that does not depend upon suspending the law or reducing individuals to their bare life. Necro-temporality is a precarious state of existing in time whereby bodies are insecure yet responsible for their own insecurity. Necro-temporality also refers to the ways certain forms of death/dying, such as dying from endemics caused by stress and insecurity, are normalized as a result of laziness, weakness, and poor character, or as just a natural part of the human condition. Thus, necro-temporality is a biopolitical technique for neoliberal governmentalities to regulate how death is to be defined, treated, and dealt with, essentially making it easier to dismiss or ignore how some populations die as natural, inevitable, or unsolvable. Furthermore, necro-temporality is a time where neoliberal subjects understand notions like "live life to the fullest" or "seize the day" as forms of resilience. Neoliberal resilience is not just about living with less. Resilience is also about living intensely in order to deal with the depravations of neoliberalism. Necro-temporality points to the ways in which populations of individuals continually live at the edge of burn out, and more and more past burn out. Insecurity, then, is in some ways valorized as a condition that pushes individuals to compete better and live more intensely.

Individuals are, in part, made insecure through racism and the racialization of bodies. Foucault suggests that racism is a precondition that allows states to kill.[31] Racism functions as a means for states to destroy whole populations as it provides methods of categorization through which these states can make live and let die. Necro-temporality highlights the ways power makes bodies vulnerable to death and exploitation by limiting time in certain spaces, forcing continual mobility, increasing time in toxic, violent and surveilled spaces, justifying these necro-temporal policies through racism, making work flexible, and displacing blame and responsibility through possessive individualism and human capital.

What I refer to in this book as the cyberpunk city or the city in general highlights an overall condition of competition that spreads over space and constitutes a general necrotic existence. To be clear, when I refer to "the city," I do not always mean a literal urban space, even though, at times, I do analyze and diagnose conditions of competition in real cities in this project. Cities in cyberpunk are also often literal cities in the sense that they are made up of urban structures, such as, densely packed architectural blocks of business and habitation, complex transportation networks, or nodes of competition (they are corporate and black-market centers). But there are also virtual spaces that exist in tandem with and develop out of the technological innovations that are constantly emerging within these urban spaces. The virtuality of city spaces in cyberpunk is not necessarily a representation of a postmodern condition, such as the end of space. There is still often a materiality that permeates the cities of cyberpunk, places where neoliberal subjects have encounters with each other, with the spaces of the cities, and with the neoliberal governmentalities that produce, engineer, and construct urban space. Thus, my goal is not to localize power forces or governmentalities within specific city spaces, to describe the space and time of cities only, or to suggest that neoliberalism is only a problem within cities, despite my repeated references to cities. Rather, I argue that the space and time of urban spaces, both in cyberpunk and in concrete cities, reveal general necrotic conditions of competition. I juxtapose cyberpunk science fiction and the narratives, materialities, and governmentalities of neoliberalism in order to denormalize, defamiliarize, and at other times familiarize this general condition of existence.

Reading the cyberpunk city as a necroscape can highlight the time and space of neoliberal governmentality in several ways. First, time in cyberpunk, or what I call necro-temporality, illustrates the ways neoliberal governmentalities in modern cities benefit from the illegal status of immigrants, and the ways state power can make some individuals permanently homeless/mobile. Even while under an illegal status, individuals are still subject to a condition of competition that makes them not only a threat to the competitive fitness of other populations, but also a danger to the healthy functioning of market competition, which according to neoliberalism, does not function naturally, and thus requires continual intervention. Making the time some people spend in certain spaces illegal limits their movement in some spaces, decreases security, and forces continual mobility, all of which contributes to their precarity. This forced continual mobility is made possible by illegal status since the illegality of their presence means that they cannot remain in one place for too long, that they cannot inhabit certain spaces for too long, and that they cannot work in one place for too long for fear that they will be caught and held subject to the law. Thus, this frequent living and movement in and through what are essentially illegal spaces makes their disappearance, exploitation, death, and experience with endemics more likely, and is thus a more precarious temporality than that which defines the dominant population. This is not to suggest, however, that the dominant population is not subject to the general conditions of competition and the precarities of its time and space. All populations are subject to accepting and calculating risk, and thus, are always potential surplus bodies, since failure to compete is a risk we all assume. Illegal status and its enforcers make some individuals increase the time they spend in dangerous spaces, for example, in the back of trucks, in the Arizona desert, in militarized zones, or in biohazardous areas. These encounters increase the likelihood of death. Illegal status is just one example of the time of surplus bodies of necro-temporality.

Second, as a necroscape, the cyberpunk city also familiarizes the ways in which neoliberal governmental rationality kills and lets bodies die through the marketization and militarization of space. For example, rather than investing in fixing current social conditions or decaying infrastructure, the economic/political/social schema of some urban spaces turn to building upwards, securitizing buildings with advanced security systems, and surveilling large portions of the city, thus leaving surplus bodies to dwell in the streets. Similar to the spatial apartheid Mike Davis describes in his analysis of Los Angeles,[32] the dominant race in cyberpunk typically secures itself, while allowing unwanted bodies to die in less secure spaces or be killed by security forces.

Finally, by juxtaposing cyberpunk literatures and neoliberal truths, we can expose the fictions of neoliberalism that identify good character or morality with individual responsibility and the rational calculation of risk, while naturalizing the idea of the entrepreneurial self. In exposing these fictions, we can make visible the "material conditions and strategic aims"[33] of neoliberal biopolitics. As a genre, cyberpunk's conceptualization of the spatial and temporal nature of a borderless market economy points to the artificiality of the market, in part by highlighting its historical contingency. Cyberpunk does this by connecting the proliferation of accelerative information technologies, such as the computer, to the development of an all-encompassing informationalized market economy. By defamiliarizing the taken-for-grantedness of the market economy and competition, cyberpunk as a genre can problematize the naturalization/ normalization of space conceived as a Neo-Darwinist jungle where the strong (individuals who are in a privileged position to capitalize and maximize their human capital) compete by weakening neoliberalism's claims to truth.

The Politics of Cyberpunk

Existing scholarship on cyberpunk has examined the ways in which this science fiction subgenre explores metaphysical questions about the nature of reality, how technology will change ways of being, and its dystopic critique of late-capitalism. In addition, scholarship on cyberpunk has offered criticisms of the conservative nature of the genre, often suggesting that it tends to re-inscribe bodies and sex within heterosexuality, while other scholarship has worked to legitimatize cyberpunk as a genre worthy of the literary canon. A dominant thread in cyberpunk scholarship has drawn from Fredric Jameson's diagnosis of postmodernism as the logic of late-capitalism, using Jameson's spatial pastiche, schizophrenic temporality, and waning of affect, along with his characterization of Jean Baudrillard's simulacrum to interpret postmodern cultural artifacts. For many cultural critics, the city of cyberpunk is thoroughly postmodern because parallels can be drawn between the cyberpunk city and Jameson's postmodern condition. Hardly any work has considered the ways in which cyberpunk examines death/dying, and scholarship on death and its relationship to the management of space and time in the context of neoliberalism, risk, and biopower has been so far very limited. In addition, much scholarship has understood the logic of cyberpunk in the context of a liberal humanist and techno-determinist schema. Thus, it has perpetuated the notion that humans and cyborgs are "tool using organisms,"[34] which occludes the possibility that tools and persons use and act upon each other.

Over the past 40 years, theorists such as Fredric Jameson, David Harvey, and Giuliana Bruno, who have worked at the intersections of postmodern theory, Marxist theory, human geography, and literary/cultural theory, have read cyberpunk in an attempt to affirm their diagnosis of the postmodern condition and to confirm the importance of space in social and cultural theory. This scholarship has examined how space has changed in a postmodern context. However, these spatial and temporal changes are too often understood as markers of postmodern aesthetics, or of the postmodern condition, rather than as indicators of neoliberal biopower, a perspective that could highlight, as I have shown above, the ways governmentality regulates populations through space and time. Thus, in much scholarship, cyberpunk becomes an indicator of a historical moment, and its critical political potentialities are limited to a Marxist critique of power. Istvan Csicsery-Ronay writes that, "[c]yberpunk is the apotheosis of the postmodern, its truest and most consistent incarnation, bar none...,"[35] and in another essay he adds that cyberpunk implies "the apotheosis of postmodernism...Hip negotiating splatter of consciousness as it slams against the hard-tech future, the techno-future of artificial immanence, where all that was once nature is simulated..."[36] Later, he describes cyberpunk's postmodern sensibility and postmodern aesthetic,[37] and he once again places cyberpunk firmly within the postmodern condition.[38] Larry McCaffery also argues that cyberpunk as the "apotheosis of postmodernism" is a "means to mirror its era's central motifs, obsessions, and desires (and render these concretely, through the dominant cultural imagery)."[39] This cyberpunk scholarship, and Bruno's and Harvey's methodologies and theoretical grounding, reaffirm normative frames about what counts as life/human. They fail to consider the power-relations that manage the spatio-temporality of cities.

Often, with a few exceptions, these theorists approach space and time in cyberpunk through a macro-political analysis, whether it is a consideration of advanced-capitalism, statehood, cultural production, or the built environment, among other concerns. Thinking about postmodernism as a mode or condition of instrumental rationality is limited as it does not offer a diagnosis of the everyday lived experiences of surplus individuals that fall outside of the normal workings of a neoliberal social system. This is not to suggest that a critique of instrumental rationality is no longer necessary. Jameson, Harvey, and Bruno, along with similar positions offered by other scholars, lament the loss of a symbolic order that affords a certain kind of agency, autonomy, and self-hood, while at the same time they accept this kind of self/individual as a form of instrumental reason. It is problematic to lament the loss of self-hood, the self, or the autonomous individual, since the individual of instrumental rationality for a long time applied only to white men, and even today often only applies to white Westerners. The individual[40] of instrumental rationality is still very much alive as a neoliberal subject. A population of individuals is made intelligible by how well these individuals subject themselves to the logic of competition. Thus, as I suggested above, the critique of the postmodern condition, as a symptom of instrumental rationality that these Marxist theory influenced scholars offer, misses the everyday endurances of surplus bodies (often not white), including chronic health problems, like obesity and diabetes, stress from systemic racism, insecurity and fear, and the requirement of constant movement to avoid detention and institutionalized violence, as bodies encounter daily the time and space of neoliberal governmentality. Jameson offers a particular way of understanding and defining power, one that does not necessarily consider the micro-politics of city life, the subtle ways systems of power promote life for some and let others die, and how space and precarity are co-produced. This anti-postmodern scholarship (anti-postmodern in the sense that the postmodern condition signals a

Introduction 13

loss of agency and fully exacerbates the problems of instrumental reason) often oversimplifies the way we can think about cities and subjectivities within cyberpunk and in real cities, and it hides the biopolitics of neoliberal discourses, such as individualized risk, self-determination, and self-responsibility.

Postmodern aesthetic readings are thus reductionist in their understanding of cyberpunk because their analysis of its literary elements, including action (living and dying), space (architecture, nature, the built environment), time, and figurative language affirms a particular interpretation of late Western Modernity that ignores other realities, governmentalities, and power/knowledge regimes, such as neoliberalism and biopolitics. Instead, a Foucauldian lens can offer insight into how we might use cyberpunk to critically examine the ways governmentality regulates space and time in order to manage death, and it can move beyond a critique of instrumental rationality that dominates much of the scholarship on cyberpunk. "Manage death" or "regulate death" refers to the ways different economic/social/cultural/political schemas kill and let individuals die, both through subtle and not so subtle modes of state killing, and through the production of subjectivities that function within and co-produce these schemas.

In the case of cyberpunk, death is often no longer the great equalizer. For Foucault, the purpose of biopower is to maintain a homeostatic population based on a norm (a manageable level of birth rates, life-expectancy, and illness of the right people, etc.). Thus, if we define neoliberalism, in part, "as a norm of life characterized by a generalized competition with others,"[41] we can understand neoliberal governmentalities as different governmental rationalities that are concerned with eliminating an excess of life/surplus bodies that fail to meet the norm of competition against others. In order to compete in cyberpunk and neoliberalism, neoliberal governmentality covertly and overtly needs to eliminate excess bodies by perpetuating a social/economic/political system that normalizes insecurity.

Postmodern Readings of Cyberpunk

Jameson and Harvey understand the production of space as inseparable from advanced/late-capitalism, and they suggest that the spaces of culture and nature are no longer separate realms. According to them, the postmodern condition is partly defined by a collapse of these semi-autonomous realms, suggesting, on the one hand, that everything can be understood as cultural, and on the other hand, that nature is an extension of multinational capitalism (pre-capitalist/pre-industrial nature is dead).[42] Furthermore, these analyses of cyberpunk point to a crisis of historicity because the notion of history as a linear progression of time with a stable subject at its core is no longer possible. Harvey adds to this debate his theory of time-space compression, which argues that the speeding up of time (again, a postmodern condition) has led to the elimination of spatial distance and barriers.

For these theorists, history (e.g., a stable human subject progressing through linear time) existed before postmodernism and has ended or mutated with the postmodern condition. But schizophrenic temporality and time-space compression privilege a Eurocentric subjectivity and temporality.[43] That is, the history/time they conceptualize is a European history, one that is "historically evolving, and geographically privileging."[44] Thinking of the subject or self as a human subject that progresses through linear time misses the ways this subject has morphed into the competitive subject of neoliberalism. In order to have a history or be a part of history, subjects must be able to participate within a linguistic order that allows them to construct a linear narrative of time. To suggest that individuals experience time as a constant present without the ability to produce a coherent biographical narrative assumes a universal conception and experience of time and history. The elimination of distance and borders through the speeding up of time applies to a particular group of people defined by race, ethnicity, and class. Some people must travel great and/or perilous distances, and they often encounter borders or other obstacles. Thus, not everyone moves fluidly from one space to another, and even those who can quickly move unencumbered through space are still subject to the general condition of competition with others, which is a precarious form of existence.

Jameson characterizes space as postmodern hyperspace, writing that it "has finally succeeded in transcending the capacities of the individual human body to locate itself, to organize its immediate surroundings perceptually, and cognitively to map its position in a mappable external world."[45] His language is immersed in the liberal tradition, drawing from liberalism's privileging of rationality, individuality, and agency. In postmodern hyperspace, individuals have lost their ability to engage in cognitive mapping. Jameson goes on to write that our inability to center the body within its built environment is analogous with "the incapacity of our minds, at least at present, to map the great global multinational and decentered communicational network in which we find ourselves caught as individual subjects."[46] Again, his concern with the loss of certain individualizing and agentic experiences reaffirms the liberal tradition and, as Tim Luke suggests, it limits the ways we can think about subjectivity. Like Luke, I suggest that we need to problematize "the accepted languages of rationality, individuality, or agency as they are used by liberalism to reproduce itself culturally, economically, and politically."[47] Not everyone is covered under the category "individual subject." Jameson's concern with the loss of our ability to engage in cognitive mapping is applicable to the traveler, the flâneur, or fit members of a dominant society, but not necessarily to surplus bodies that fail to compete in a market economy or pose a threat to the dominant group's ability to compete. In his examination of postmodern space, Jameson laments the loss of our ability to promenade on our own or engage in the narrative stroll because the escalator does the walking/traveling for us. But promenading and strolling were always part of a bourgeois fantasy, not necessarily a universal experience.

According to Jameson, cyberpunk is a "degraded attempt...to think the impossible totality of the contemporary world system."[48] That is, the "immense communicational and computer networks"[49] of cyberpunk are a faulty representation of a technology that is in itself a faulty representation of reality. Cyberpunk is another example of the weaker productions characteristic of postmodernism. While I agree that cyberpunk expresses "transnational corporate realities," I would not go so far as to call it an expression of global paranoia, nor would I consider its computer technology as a machine of reproduction. It is more interesting and perhaps more accurate to think about the computer in terms of its productive and reproductive capabilities. The machine of reproduction does not account for the unpredictability, ingenuity, and liveliness of machine learning. Thus, cyberpunk is more than a narrative "about the processes of reproduction."[50]

As I suggested above, much of the scholarship that came after Jameson's 1984 works "Postmodernism, or, The Cultural Logic of Late Capitalism" and "Postmodernism and Consumer Society" was influenced by his analysis of postmodernism and his diagnosis of the postmodern condition. For example, in his reading of *Blade Runner*, which is generally considered a cyberpunk film, Harvey writes that time and space represent "instantaneous global communications."[51] Harvey uses Jameson's understanding of temporality when he describes how replicants exist in time, calling it a "schizophrenic rush of time." And he writes that they "also move across a breadth of space with a fluidity that gains them an immense fund of experience."[52] Harvey combines this time-space compression with Jameson's schizophrenic temporality. But in doing so, Harvey fails to capture the different borders that replicants and some non-replicants have to traverse or violate in order to move from one space to another. These individuals are blocked from entering Earth, and, therefore, must enter illegally; they are blocked from entering the upper levels of the corporate world, and, therefore, must enter illegally; they move quickly through the city when they can, but not because technology has eliminated distance and time. Instead, they move quickly because they are being hunted by a police/corporate state. The city is too crowded to suggest that anyone can move fluidly, except perhaps for the blade runners, who can avoid the streets via flying cars. This fluid and fast movement of information is also not necessarily applicable to the everyday movement of bodies in the city.

In an anthology on cyberculture, Scott Bukatman borrows from Jameson and Harvey in his book *Terminal Identity* to conceptualize space in cyberpunk and contemporary urban reality. He argues that contemporary urban space is boundless, directionless, and motionless. Bukatman writes that the "urban territory is marked by an infinity of space, a multiplicity of surfaces: time is displaced within a field of inaction and, ultimately, inertia as the city, the universe, circles back upon itself in a closed feedback loop...the city-state has become the *cybernetic state*."[53] Bukatman's analysis of the city without a center and limits is overstated since it fails to account for the encounters different individuals have with territorial borders. The space of cyberpunk is both geographical and non-geographical; there is continual tension between bodies moving in real physical space and virtual bodies in the mental geography of cyberspace. Bukatman's essay focuses primarily on the spatial similarities between cyberspace and urban space, highlighting an aesthetics of space. Like Jameson and Harvey, he emphasizes and diagnoses the architectural shifts of the postmodern condition, including changes in style, size, and location, along with the changing landscape/cityscapes of urban spaces. Again, this is an approach that, as I suggested above, ignores the movements of bodies in space and the power/knowledge regimes that regulate them.

Aims and Chapters

Aims

One aim of this book is to make strange seemingly non-neoliberal notions, everyday practices, and material conditions, such as self-cultivation, biohacking, biopolitics, cyborgs, science fiction, postmodern, and exercise. By doing so, I highlight the ways in which neoliberalism perpetuates itself in not so obvious ways. I am *not* trying to defamiliarize the term neoliberalism itself or, for that matter, to defamiliarize the political endeavors of neoliberalism. Here, I do not focus primarily on the political machinations of neoliberalism because I am more interested in neoliberalism's micro-practices, in the minutiae of neoliberalism as an ideology, a practice of power, and a form of life that often invests populations of individuals. I focus on the minutiae of neoliberalism, further brought into relief by juxtapositions with cyberpunk forms. Indeed, too often, studies of neoliberalism are state-centered, and as such, they miss or ignore the subtle, insidious, and de-centered ways by which neoliberalism functions.

Another aim of this text is to examine intensity and resilience as dominant logics of neoliberalism. Specifically, my position is that intensity and resilience highlight the ways in which biopolitics functions under neoliberalism, namely that making live, or biopolitics, can now be understood through an incitement to live intensely in order to remain resilient.

In order to live and make live in ever increasing insecure and harsh conditions, neoliberal governmentalities and subjects operate through a logic of intensity. That is, neoliberal biopolitics does not only function by making populations live by norms. Rather, the logic of intensity dictates the need to exceed norms: to live intensely. For some populations, this means living always on the edge of burnout and for others it means living like they are on the edge of burnout. And, the logic of intensity eliminates certain populations that fail to avoid burnout (e.g., the inability to compete) since it operates within harsh conditions, effectively displacing responsibility for these harsh conditions away from neoliberal governmentalities onto individuals.

I approach this theoretical examination of the logic of intensity by focusing on the co-productive relationship between art and neoliberalism and on the everyday micro-practices of living intensely and precarious survival. Again, I move away from more traditional disciplinary aesthetic methods of analyzing power and urban systems, such as interpretation and representation, to problematize the biopolitical present, which defines and controls life through an integration of economic calculation and economic rationalities into life and bodies.

Chapters

The study that follows is comprised of five chapters that are generally guided by four themes: biopolitics, intensification, resilience, and accelerationism. The first chapter examines the political possibilities of science fiction by treating cyberpunk as neoliberal politics. I show how cyberpunk's literary and filmic techniques, including amplification and cyborganization, can make strange the neoliberal biopolitical present. Then, Chapter 2 continues with this process of making strange as it problematizes notions of self-cultivation and neoliberal health that, as Thomas Lemke points out, neoliberalism accepts as the way it has always been. Thus, Chapter 2 examines a concrete example of a cyberpunk neoliberal subject (the self-monitoring cyborg) to show that, while self-cultivation is no longer possible under neoliberal capitalism, neoliberalism still capitalizes on notions of self-cultivation. The self-monitoring cyborg instrumentalizes self-cultivation as a way to increase its human capital. These everyday self-monitoring practices are just one way the neoliberal subject subjectivizes itself.

In Chapter 3, I briefly shift more overtly to biopolitics. I look at how recent scholarship on biopolitics has argued that Foucault's analytic for knowing power can no longer fully capture the ways life is governed and made an object of power. Rather than advocating for moving beyond biopolitics, this chapter argues that neoliberal biopolitics can still be understood in terms of Foucault's analytic, and that perhaps we need to disentangle Foucault's work from Achille Mbembe's "Necropolitics." The suggestion here is that we miss something about the way neoliberalism and biopower function if the focus of scholarship in biopolitics is on overt and massive-scale state killing. Focusing on overt and massive-scale state killing may suggest that these modes of killing are the only form of state killing, and thus, in some ways, further neoliberal realities where individuals are to blame for their endemic and economic suffering.

Following Chapter 3, I move from Foucault's biopolitics to intensification to highlight the fact that biopower makes live through neoliberal resilience, which is largely governed by a logic of intensity. Again, part of the goal of neoliberal governmentalities is to place responsibility on individuals. Thus, one way to do this is to incite individuals to live intensely so that they will want to live and accept the responsibility of living in insecure conditions. Chapter 4 develops further the themes of intensity and resilience as it presents a position on space and time that is understood in the context of neoliberalism. It does this by considering the cyberpunk film *Blade Runner* as a productive force in reproducing and perpetuating notions of neoliberal resilience. Chapter 4 argues that the time of neoliberalism can be characterized in terms of necro-temporalities since the resilience that I consider needs individuals to want to live insecurely so that they can live more intensely always at the edge of burn out. In Chapter 5, I return to biopolitics, intensity, and resilience, and connect all of these themes to accelerationism as I engage with biohacker discourses. The biohacker is, in part, an intensification of the self-monitoring cyborg. Thus, I argue that biohackers are neoliberal subjects in order to show that accelerationism is in the end another form of resilience. To endure the accelerated horrors of neoliberalism and capitalism, biohackers live the most intense lives. The "post-capital" future of accelerationism remains a neoliberal utopia with nightmarish possibilities. Finally, the conclusion returns to some of the core considerations of this study. First, it looks back at the ways in which the book defamiliarizes neoliberalism's reality construction. It then briefly considers how this study rethinks biopolitics in its neoliberal form. And, finally, this conclusion returns to thinking about the logic of intensity as a form of neoliberal resilience as it weaves through the narrative flow of this study.

A Note on Organization

My chapter progression throughout the text is designed to mirror the logic of intensification that is so prolific within neoliberalism. Under neoliberalism, individualism is intensified to the point where individuals want and accept harsh living conditions, and these individuals further pride themselves on the ability to remain resilient in the face of environmental depravations. According to the logic of intensity, the steps one takes to remain resilient are never enough. Resilience, while often a goal, is itself not enough to remain competitive. The ability to bounce back quickly, to endure, and to never give up all point to some level of individual weakness. That one has to bounce back, or to endure, or to never give up shows that the individual is by definition susceptible or vulnerable to something, whether it is an illness, stress, pollution, etc. Thus, my movement in the text from self-monitoring cyborgs to biohackers highlights the ways that neoliberal governmentalities and neoliberal subjects intensify resilience and self-cultivation to superhuman or seemingly more than human proportions. I begin with my chapter on cyberpunk genre techniques, such as amplification (itself a type of intensification) and cyborganization, to set the foundations for how the study defamiliarizes neoliberalism. Chapter 1 establishes a way of thinking strangely about competition, responsibility, individualism, and risk, an approach that carries throughout the study. Chapter 2 introduces one kind of neoliberal subject, the self-monitoring cyborg that, as it gains more human capital, later becomes the biohacker that we see in Chapter 5. And I include midway through the text a chapter that disentangles Foucault's biopolitics from Mbembe's necropolitics to bring into relief the necrotic conditions that biohackers face. Chapter 3, placed as it is in the study, offers additional ways of thinking strangely about neoliberalism and how neoliberal states kill populations. It shows that states can kill populations without directly causing their deaths, and that in the context of neoliberalism and its biopolitics, individuals want insecure, risky, dangerous, or harsh living conditions in order to live intensely. It is placed in the middle of the study to highlight, in a biopolitical context that is about living intensely, necrotic conditions such as living on the edge of burnout or competing to live that allow for a shift from a less intense neoliberal subject (the self-monitoring cyborg) to a much more intensified neoliberal subject (the biohacker). In other words, Chapter 3 seeks to think strangely about biopolitics in order to highlight some of the ways in which neoliberalism deresponsibilizes its governmental rationalities, including the logic of intensity and accelerationism, both of which are a productive force for the biohacker. By mirroring neoliberal intensification, my goal is to make it easier to see one way that neoliberalism constructs its reality, to bring intensification to our immediate attention, and to show that the logic of intensity is specific to neoliberalism. It then becomes harder to accept living intensely and wanting to live in insecure conditions to prove one's resilience as just the way it is, as just the way it has always been, or as the only way it will be. This means that, with each chapter, living intensely intensifies and becomes more and more apparent as a mode of living within neoliberalism.

Notes

1. Thomas Lemke, "The Risks of Security: Liberalism, Biopolitics, and Fear," in *The Government of Life: Foucault, Biopolitics, and Neoliberalism*, eds. Vanessa Lemm and Miguel Vatter (New York: Fordham University Press, 2014), 68.
2. Gordon Hull, "Biopolitics Is Not (Primarily) About Life: On Biopolitics, Neoliberalism, and Families," *The Journal of Speculative Philosophy* 27, no. 3 (2013).
3. For this project, I am primarily concerned with neoliberal governmentality within the United States.
4. Loïc Wacquant, *Punishing the Poor: The Neoliberal Government of Social Insecurity* (Durham: Duke University Press, 2009).
5. Stephen Graham, *Cities Under Siege: The New Military Urbanism* (New York: Verso, 2011).
6. Thomas Lemke, "Foucault, Governmentality, Critique" (presentation, The Rethinking Marxism Conference, University of Amherst (MA), September 21–24, 2000): 13.
7. Michel Foucault, *History of Sexuality: An Introduction*, Volume 1 (New York: Vintage, 1990), 72.
8. Lemke, "Foucault Governmentality," 13.
9. Cyberpunk is a subgenre of science fiction that often follows the risk taking adventures of an antiauthoritarian protagonist in a high tech dystopian future. In many ways cyberpunk takes some of the tropes of detective noir (e.g., an anti-hero protagonist, set in the seedy underbelly of cities, often involves a faceless figure of power that is too big to take down, femme fatale figures, etc.) and places them within the context of a dystopian future dominated by the effects of late-capitalism. The anti-hero of detective noir becomes the cyber cowboy, the seedy underbelly of cities now contains black markets for synthetic flesh and drugs, and the faceless power becomes a corporation.
10. Ibid., 154.
11. Michel Foucault, *The Birth of Biopolitics: Lectures at the Collège de France, 1978–79*, translated by Graham Burchell (New York: Palgrave MacMillan, 2008).
12. Michael Shapiro, *The Time of the City: Politics, philosophy, and genre* (New York: Routledge, 2010), 4
13. Steven Shaviro, *The Cinematic Body* (Minneapolis: University of Minnesota Press, 1993).
14. These political and economic genres include the political and economic aims of neoliberal governmentalities and the underlying political and economic philosophies, such as, the condition of competition, that subtends neoliberalism.
15. Ibid., 4.
16. Gilles Deleuze, *Cinema 2* trans. by Hugh Tomlinson and Robert Galeta (Minneapolis: University of Minnesota Press, 1989), 280.
17. Elizabeth Povinelli, *Geontologies: A Requiem to Late Liberalism* (Durham: Duke University Press, 2016), 4.
18. Elizabeth Povinelli, *Geontologies: A Requiem to Late Liberalism* (Durham: Duke University Press, 2016), 134.
19. See Elizabeth Povinelli, *Geontologies: A Requiem to Late Liberalism* (Durham: Duke University Press, 2016); François Debrix and Alexander Barder, *Beyond Biopolitics: Theory, Violence, and Horror in World Politics* (New York: Routledge, 2012).

20. Elizabeth Povinelli, interview by Mathew Coleman and Kathryn Yusoff, "An Interview with Elizabeth Povinelli: Geontopower, Biopolitics and the Anthropocene," *Theory, Culture & Society* 34, no. 2–3 (2017): 170.
21. Ibid., 170.
22. Bruno Amable, "Morals and Politics in the Ideology of Neo-liberalism," *Socio-Economic Review* (2011): 7.
23. Ibid., 5
24. Stephen Graham, *Cities Under Siege: The New Military Urbanism* (London: Verso, 2010).
25. Ibid., xiii, xxi.
26. Achille Mbembe. "Necropolitics," *Public Culture* 15, no. 1 (2003): 40.
27. Ibid., 103.
28. Ibid., 103.
29. Giorgio Agamben, *Homo Sacer: Sovereign Power and Bare Life*, translated by Daniel Heller-Roazen (Stanford: Stanford University Press, 1998).
30. Carl Schmitt, *Dictatorship* (Malden: Polity Press, 2014).
31. Michel Foucault, "*Society Must Be Defended*": Lectures at the Collège de France, 1975–76, trans. David Macey, New York: Picador, 1997), 262–263.
32. Mike Davis, *City of Quartz* (New York: Vintage, 1992).
33. Lemke, "The Risks of Security," 68.
34. Tim Luke, "Liberal Society and Cyborg Subjectivity: The Politics of Environments, Bodies and Nature," *Alternatives: Global, Local, Political* 21, no. 1 (1996): 8.
35. Larry McCaffery et al., "Cyberpunk Forum/Symposium," *Mississippi Review* 16, no. 2 (University of Southern Mississippi, 1988), 27.
36. Istvan Csicsery-Ronay, "Cyberpunk and Neuromanticism," *Mississippi Review* 16, no. 2/3 (1988): 266.
37. Ibid., 269.
38. Ibid., 271.
39. Larry McCaffery, "The Desert of the Real: The Cyberpunk Controversy," *Mississippi Review* 16.2/3 (University of Southern Mississippi, 1988): 8.
40. This individual instrumentalizes the world as a way to ensure self-preservation. The norm for self-preservation is now competition.
41. Bruno Amable, "Morals and Politics," 7, 9.
42. Fredric Jameson, *Postmodernism, or, The Cultural Logic of Late Capitalism* (Durham: Duke University Press, 1991), 49.
43. See Michael Shapiro, "The now time(s) of the global city: Displacing Hegel's geopolitical narrative," *The Time of the City: Politics, philosophy and genre* (New York: Routledge, 2010), 25–45; see also Aimé Césaire, *Discourse on Colonialism*, trans. Joan Pinkham (New York: Monthly Review Press, 1972).
44. Shapiro, *The Time of the City*, 27.
45. Jameson, *Postmodernism*, 44.
46. Ibid., 44.
47. Luke, "Liberal Society," 2.
48. Jameson, *Postmodernism*, 38.
49. Ibid., 37.
50. Ibid., 37.
51. Harvey, 309.
52. Ibid., 309.
53. Scott Bukatman, *Terminal Identity: The Virtual Subject in Postmodern Science Fiction* (Durham: Duke University Press, 1993), 128.

Bibliography

Agamben, Giorgio. *Homo Sacer: Sovereign Power and Bare Life*. Translated by Daniel Heller-Roazen. Stanford: Stanford University Press, 1998.

Amable, Bruno. "Morals and Politics in the Ideology of Neoliberalism." *Socio-Economic Review* 9 (2011): 3–30.

Bukatman, Scott. *Terminal Identity: The Virtual Subject in Postmodern Science Fiction*. Durham: Duke University Press, 1993.

Césaire, Aimé. *Discourse on Colonialism*. Translated by Joan Pinkham. New York: Monthly Review Press, 1972.

Csicsery-Ronay, Istvan. "Cyberpunk and Neuromanticism," *Mississippi Review* 16, no. 2/3 (1988): 266–278.

Deleuze, Gilles. *Cinema 2*. Translated by Hugh Tomlinson and Robert Galeta. Minneapolis: University of Minnesota Press, 1989.

Foucault, Michel. *The Birth of Biopolitics: Lectures at the Collège de France, 1978–79*. Translated by Graham Burchell. New York: Palgrave MacMillan, 2008.

—. *History of Sexuality, Volume 1: An Introduction*. New York: Vintage, 1990.

—. *"Society Must Be Defended": Lectures at the Collège de France, 1975–76*. Translated by David Macey. New York: Picador, 1997.

Graham, Stephen. *Cities Under Siege: The New Military Urbanism*. New York: Verso, 2011.

Hull, Gordon. "Biopolitics Is Not (Primarily) About Life: On Biopolitics, Neoliberalism, and Families." *The Journal of Speculative Philosophy* 27, no. 3 (2013): 322–335.

Jameson, Fredric. *Postmodernism, or, The Cultural Logic of Late Capitalism*. Durham: Duke University Press, 1991.

Lemke, Thomas. "The Risks of Security: Liberalism, Biopolitics, and Fear," in *The Government of Life: Foucault, Biopolitics, and Neoliberalism*, edited by Vanessa Lemm and Miguel Vatter, New York: Fordham University Press, 2014. 59–76.

—. "Foucault, Governmentality, Critique." Presentation at The Rethinking Marxism Conference, University of Amherst (MA), September 21–24, 2000.

Luke, Tim. "Liberal Society and Cyborg Subjectivity: The Politics of Environments, Bodies and Nature." *Alternatives: Global, Local, Political* 21, no. 1 (1996): 1–30.

McCaffery, Larry et al. "Cyberpunk Forum/Symposium," *Mississippi Review* 16, no. 2. (1988): 16–65.

—. "The Desert of the Real: The Cyberpunk Controversy," *Mississippi Review* 16, no. 2/3. (1988): 7–15.

Povinelli, Elizabeth. *Geontologies: A Requiem to Late Liberalism*. Durham: Duke University Press, 2016.

Schmitt, Carl. *Dictatorship*. Malden: Polity Press, 2014.

Shapiro, Michael. *The Time of the City: Politics, Philosophy, and Genre*. New York: Routledge, 2010.

Shaviro, Steven. *The Cinematic Body*. Minneapolis: University of Minnesota Press, 1993.

Wacquant, Loïc. *Punishing the Poor: The Neoliberal Government of Social Insecurity*. Durham: Duke University Press, 2009.

1 The Neoliberal Science Fictions of Cyberpunk

Introduction

As I suggested in the Introduction, this project moves away from more traditional disciplinary aesthetic methods of analyzing literature and film, such as interpretation and representation. In this chapter, I wish to problematize the biopolitical present by weaving in and out of an analysis of the narratives, discourses, and spatio-temporalities of cyberpunk and neoliberalism. By moving/writing/thinking in and out of literary and political/philosophical genres, I seek to produce a series of epistemological interferences within these genres/disciplines, and thus, to disrupt the conceptual and lived biopolitical status-quo of neoliberalism. The goal is to open the door for discomfort with and a critical awareness of the necrotic conditions of competition by highlighting the narrative/fictive nature of the political, moral, and economic theories/practices of neoliberalism. These interferences, as Shapiro argues, help to challenge "the already conceptually invested political status-quo,"[1] and I would suggest the economic and social status-quo too, since it is the case that under neoliberalism "good" everyday life functions through an economic logic.

Cyberpunk can help to bracket and make conscious the necrotic conditions of competition through its spatialization and temporalization of neoliberal biopolitics, security, and corporate capitalism. As a genre, cyberpunk can make familiar the conditions of competition, such as the marketization of space and the production of responsibilized subjectivities, via notions of amplification and cyborganization. The critical possibilities of cyberpunk emerge out of its amplification and cyborganization of the spaces of neoliberal biopolitics and corporate capitalism as it is these genre techniques that weaken the distinction between fiction and the narratives that permeate the everyday.

The Affirmative Speculation of Cyberpunk Science Fiction

In his essay "The Co-Existence of Cyborgs, Humachines and Environments in Postmodernity: Getting Over the End of Nature," Timothy Luke comments on the ways science fiction makes visible the science facts of a cyborganized environment. These cyborganizing/ed sites and structures highlight the "'environmental spaces' of transnational capitalist exchange where cyborg beings are cyborganizing/ed..."[2] For Luke, the figure of the cyborg acts as an analytic for examining the self-evident truths of late-modern liberalism: the notion of a pristine nature independent of the human/cultural/political/economic realm, the preservation of the free individual as the central goal of politics and the state, and the free market as a means of protecting the individual. As a denatured and dehumanized being, the cyborg problematizes liberalism by offering an alternative politics with different subject categories, environments, spaces, and temporalities. In other words, liberalism does not recognize the science fact of cyborganized environments and ontologies, nor can it see the cyborg subjectivities "just beneath the surface of liberal society."[3] Liberal political theories miss the "modernization project of world-systemic capitalism,"[4] which denatures and dehumanizes[5] as it seeks to integrate all areas of life into "global networks of exchange."[6]

Luke's cyborg theory shows how science fiction can "bracket," "engage," and "make conscious" the shortcomings of liberal political theory and the workings of power, along with "demystifying" a denaturalist reality. For Luke, science fictions "...are (ab)useful, illusions for (re)inventing our imagination of power, economy, and culture in an environment materially built into and out of scientific facts."[7] Similar to the ways Luke's cyborg myth/science fiction familiarize liberal categories and liberal politics, cyberpunk can familiarize the spatial and temporal logic of neoliberalism.

As we saw in the Introduction, Jameson argues that science fiction does not give us visions of the future. Instead, science fiction has a certain kind of realism that "defamiliarizes and restructures our experience of our own present," and it does "so in specific ways distinct from all other forms of defamiliarization."[8] Jameson argues that science fiction is an "elaborate strategy of indirection," one that offers a gaze through which we can view and critique the present. For Scott Bukatman, "[s]cience fiction...is grounded in the new 'intolerable spaces' of technological culture and the narrative exists to permit that in a manner now susceptible to human perception, comprehension, and intervention..."[9] Science fiction provides a conceptual frame.[10] This chapter addresses the ways in which cyberpunk as a genre can denaturalize and familiarize the present, and in so doing, problematize neoliberal notions by highlighting their narrative qualities. I focus on amplification and cyborganization as genre conventions of cyberpunk science fiction, and I show how these two genre conventions are key to how cyberpunk denaturalizes and familiarizes the present. Amplification and cyborganization offer possibilities for a critical awareness that weakens the distinctions between science fiction and neoliberal narratives that can then make the self-evident nature of these narratives into a problem. The neoliberal narratives that I will address are those that insist on the fact that there is a clean separation between the economic and the social, that a subject's character and personhood are defined by individualized responsibility and calculations of risk, and that the free market is a natural phenomenon.

Amplification

Melinda Cooper writes that the capitalist promise, and I think by extension the neoliberal promise too, "is counterbalanced by willful deprivation, its plenitude of possible futures counteractualized as an impoverished, devastated present, always poised on the verge of depletion."[11] As a mode of neoliberal accelerationism,[12] cyberpunk pushes this willful deprivation, as well as the impoverished and devastated present, into some visions of the future. Cyberpunk plays with accelerationism's argument that "the only way out is the way through,"[13] by offering a vision of the future past the "long-term, slow-motion catastrophe"[14] of the perpetual crises that undergird generalized conditions of competition. It highlights the contradictions that function within neoliberalism (the free market is natural/competition is not natural, the economic and social are separate/the economic and the social are not separate). Neoliberalism benefits from lingering narratives of liberalism (the economic and social are separate, the free market is a natural phenomenon, competition is natural) because these narratives hide or make natural/taken-for-granted the ways neoliberal governmentalities manage their subjects in order to intensify processes of competition. As Michel Foucault highlights in "*The Birth of Biopolitics*," unlike liberalism, neoliberalism does not view competition as a process that occurs naturally.[15] Rather neoliberalism must work at producing and maintaining the conditions of competition. Life, subjects, society, and politics become intelligible through market competition. The neoliberal governmentalities that shape and make legible the relationships, interactions, and experiences of individuals in terms of competition are unwilling to recognize these contradictions and the insecurity that emerges out of these relationships/interactions/experiences. As readers interact with cyberpunk, the responsibilized self, the naturalized free market, and the insecurities that develop out of the conditions of competition come into focus as they unfold within the cyberpunk narrative and through its depiction of bodily encounters in city spaces.

According to Loïc Wacquant, "neoliberal ideology in economic matters rests on an impermeable separation between the economic (supposedly governed by the neutral, fluid, and efficient mechanism of the market) and the social (inhabited by the unpredictable arbitrariness of powers and passions)."[16] The separation Wacquant describes hides what Foucault explains is fundamental to neoliberalism: neoliberalism generalizes an analysis of the market economy beyond the economic field as a means of intelligibility.[17] As I suggested above, neoliberalism understands the relationships/interactions/experiences of individuals in terms of an "economic game of competition."[18] In such a game, the city is often seen as a space for subjects to enhance and utilize their human capital as they compete for jobs. Thus, when people fail to compete, this failure is seen not as a systemic effect, but rather as an individual weakness of character or a subject malfunction. The suggestion that their failures are systemic is often interpreted as de-mobilizing and de-responsibilizing. Wacquant suggests that the "virile rhetoric of personal uprightness and responsibility" is "tailor-made for deflecting attention away" from the state's inability, or perhaps unwillingness, to intervene.[19] And yet, it is not just neoliberal subjects that are made intelligible through the logic of competition. It is also the case that a logic of competition makes certain governmentalities intelligible, since as Foucault suggests, under neoliberalism, there is a permanent "neoliberal market" criticism (e.g., "scrutinizing every action of the public authorities in terms of the game of supply and demand, in terms of efficiency...and the cost of intervention"[20]) of different governmentalities that are not filtered through competition and an economic grid.[21]

Cyberpunk complicates the myth of the separation between the economic and the social by amplifying to excess the generalization of market economies beyond the economic field. The economic permeates the social, political, and spatial fields of cyberpunk. William Gibson's novel *Neuromancer* offers a good example of this bleeding of the economic into other spheres. When the protagonist of the novel, Case, finds himself without money, he uses whatever means possible to reenter the economic game of competition:

> At first, finding himself alone in Chiba, with little money and less hope of finding a cure, he'd gone into a kind of terminal overdrive, hustling fresh capital with a cold intensity that had seemed to belong to someone else. In the first month, he'd killed two men and a woman over sums that a year before would have seemed ludicrous. Ninsei wore him down until the street itself came to seem the externalization of some death wish.[22]

This passage illustrates Case's attempt to replenish his human capital as he commits violent acts. He is in some sense justified in committing these acts by the logic of neoliberal temporalities in Chiba city. The spatial order of Chiba city is dominated by economic competition, which frames many of the social interactions that occur within the city, including murder. *Neuromancer* illustrates one way in which the cyberpunk genre spatializes a neoliberal reasoning that "would praise and privilege the 'rational' logic of unscrupulous, self-preserving, self-determining corporations over" less "rational" actors.[23] We are expected to sympathize with Case's rational logic as he is the protagonist and individualistic hero of the text. Yet, at the same time, Case's "choices" bring into stark relief the monstrosity of the neoliberal subject in excess.

A little later, the narrator describes the temporal logic of a large section of Chiba city:

> Night city was like a deranged experiment in social Darwinism, designed by a bored researcher who kept one thumb permanently on the fast-forward button. Stop hustling and you sank without a trace, but move a little too swiftly and you'd break the fragile surface tension of the black market: either way, you were gone, with nothing left of you but some vague memory...Biz here was a constant subliminal hum, and death the accepted punishment for laziness, carelessness, lack of grace, the failure to heed the demands of an intricate protocol.[24]

Neuromancer's narrator hints at neoliberal values that underlie the spatial order, not only of Gibson's text but also of cyberpunk in general. Death in Night City, whether from laziness, carelessness, or a lack of grace, is understood as a failure to maximize one's human capital. Cyberpunk thus can highlight neoliberalism's need for insecurity to "induce foresight and prudence."[25] It shows that neoliberal biopolitics[26] operates through a fundamental insecurity that "present[s] society as an 'exposed community'... Coping with fear becomes a problem of individual psychology or a medical issue, while the material conditions and the strategic aims of the production of fear remain invisible."[27] Insecurity is not simply an undesirable effect of freedom. Rather, insecurity is, as Thomas Lemke contends, an "essential and positive condition" of liberal and neoliberal freedom. Neoliberalism "nurtures danger, it subjects danger to an economic calculus...." It "must never fix security."[28] Indeed, it must cultivate it. It must make insecurity thrive.

Cyberpunk sets its totalizing gaze on the city or suburb as the free play of the market, and through this gaze, it offers a moment of focus on the end results of disassociating social causes from responsibility. Through amplification, a text like Neal Stephenson's *Snow Crash* can highlight the temporal logic that emerges out of neoliberalism, or what I called necro-temporalities in the Introduction. In *Snow Crash*, people live in burbclaves, which are suburbs that have become city-states with their own "constitution, border, laws, cops, everything."[29] Burbclaves are securitized spaces (conglomerations of houses, schools, medical facilities, and strip malls) that are surveilled by privatized police forces for hire and through surveillance technologies. There are no free spaces in *Snow Crash*. Individuals must be vetted by security systems before they can enter burbclaves, and they must have money to justify their presence in city spaces. Reflecting primarily on Los Angeles, Mike Davis writes that "[a]nyone who has tried to take a stroll at dusk through a strange neighborhood patrolled by armed security guards and signposted with death threats quickly realizes how merely notional, if not utterly obsolete, is the old idea of the 'freedom of the city.'"[30] Still, the liberal notion that individuals are free to move and make choices, at least for some individuals, persists in a world where the only option for survival is to compete. That is, freedom and individuality are tied to choice and mobility through machines that maximize speed in terms of time and space. Freedom in cyberpunk's burbclaves and cities should seem familiar to neoliberal subjects, since it is also the case that neoliberal subjects are free if they can choose the fastest car and the fastest digital access in order to maximize their human capital.

Under the conditions of competition that shape and produce the time and space of cities and burbclaves, individuals must alter their bodies through bio-enhancements in order to maximize their human capital and avoid sinking without a trace.[31] Furthermore, under a neoliberal logic, the desire for these enhancements is an individual choice rather than a result of insecure material conditions. The separation between the economic and the social ignores these insecure material conditions and other systemic problems. But cyberpunk does not ignore these conditions. Instead, it visually spatializes the insecure spaces of neoliberalism. In a similar vein, Benjamin Noys writes that cyberpunk captures "the thrill and threat of *materialization*...The threat is from bad tech, bad surgery, and falling behind the accelerated race to the future."[32] The insecure cities of cyberpunk are the flipside of neoliberal economic rationality since they make visible the material conditions of insecurity. They also reveal blatantly that these conditions/spaces of insecurity are crucial to neoliberalism and neoliberal subjectivities. Thus, the amplification of the economization of the social to excess has the potential to shock the reader/viewer into a critical awareness of their neoliberal present as they are forced to turn a critical eye to their own spaces.

Cyborganization

I define cyborganization as the proliferation of bodies that are neither human/organic nor non-human/non-organic. Through these integrated bodies, cyborganization reflects an underlying drive to maximize human capital through hybridic bodily formations. Cyborganization, in cyberpunk in particular, points to the non-radical reality of cyborg forms as they are motivated by competition and subjectivized by neoliberalism.[33] Cyborganization also often points to how the cyborg stands in for the racialized or subaltern other. Even within the context of these cyborganized futures, subjectivity and the rights that come with it are dependent on definitions of humanness. That is, cyborgs may force people within cyberpunk worlds to rethink what it means to be human, but this rethinking does not necessarily eliminate the category of human. Being human is still a hierarchical category that others must meet in order to be included in the social and political order, which is why this project is a biopolitical one. It is about a certain conception, production, and subjectivation of human life.

The successful human subject in cyberpunk is the human of the liberal humanist tradition and at the same time it is more than human (this subject defies death, pushes beauty norms to excess, maximizes strength and agility, makes real the cartesian dream of the separation of mind from body). It is a hyperhuman (although I will suggest later in this chapter that the hierarchical structures in the original *Blade Runner* offer an exception to the hyperhuman rule). This hyperhuman has a higher capability of competing in a neoliberal market. Thus, biopolitics, as it functions in cyberpunk, points to the neoliberal governmentalities that let die or kill surplus bodies (subjects that are not quite hyperhuman or are unintelligible to neoliberal governmentalities) through criminalization, biological racism, and insecurity as they work to perpetuate conditions of competition.[34] I am using Foucault's understanding of racism here. According to Foucault, racism within biopolitics is a biological racism that separates a population from another population that is deemed a threat to the overall health of the valuable/dominant population. Biological racism is a means of justifying killing in the context of biopolitics. In the case of another cyberpunk novel, Richard K. Morgan's *Altered Carbon*, the wealthy have access to subtler forms of bodily enhancement in order to maintain perfected human forms. With enough money, these wealthy individuals can build large production/housing centers that produce and preserve their genetically and chemically enhanced "sleeves."[35] Their value is tied to their wealth as it is through their money that they attain human status, and it is also this wealth that affords them political clout in defining their human status.

The film *Blade Runner*[36] exemplifies the critical possibilities of cyborganization too. *Blade Runner* spatializes and temporalizes what I call necro-temporalities and necro-spaces. My use of necro-temporalities and necro-spaces considers the ways in which neoliberal subjects, in general, are made to live by competition and to accept the risks and responsibility of their choices. In other words, I consider the ways in which we are all made more insecure in neoliberalism in later chapters. Here, though, I briefly address what Lisa Marie Cacho describes as spaces of living death as I think this notion is indicative of the literature on neoliberal spatio-temporalities. Lisa Marie Cacho, among others, is largely concerned with the ways in which racism in the context of neoliberalism and corporate capital make racialized populations more vulnerable. These spaces of living death exist within and are perpetuated primarily by corporate capital and neoliberalism, as well as their agents (e.g., blade runners). Borrowing from Sharon Holland's work,[37] Cacho argues that "racism is a killing abstraction,"[38] which references "the ways in which racialized populations are made unduly vulnerable by global capitalism and neoliberal restructuring."[39] In *Blade Runner*, corporate capitalism has terraformed the earth[40] to produce a denatured environment where its economic logic can thrive. Time is accelerated in this world where, in order to succeed, subjects must always be mobile, whether through high speed vehicles, visual technologies, or information technologies. However, this accelerated mobility is differentially experienced. That is, blade runners, humans, and corporate elites can move vertically and horizontally at high speeds; however, once replicants re-enter earth, their mobility is for the most part limited to the horizontal movement through city streets. As I suggested above, the hierarchical structures that exist within *Blade Runner* consider hyperhumans (replicants) a threat to the competitive success of the dominant population. Thus, *Blade Runner* may be seen as an exception to the hyperhuman rule in terms of who defines and regulates successful neoliberal subjects. In the case of *Blade Runner*, it is not hyperhumans. Cyberpunk visualizes, both through the conceptual realm of novels and the image realm of films, the acceleration of time and denatured environments of an informationalized present. This capitalist and neoliberal restructuring of space also familiarizes the spaces of living death that replicants and other surplus bodies must face once they enter the city. Replicants in *Blade Runner* are criminalized cyborg bodies. Their criminalization/racialization also determines their movement through the city,[41] including the time they can spend in certain spaces, and the kind of spaces they can encounter.

Blade Runner thus captures the exclusionary practices that make possible the juridical and economic systems of neoliberalism's present. Just as insecurity is an essential and positive condition of neoliberalism, exclusion is an essential and positive condition of the law, which means that the "law is dependent upon the permanence of certain groups' criminalization."[42] Replicants are one of the populations that are "ineligible for personhood,"[43] as they are held to the law but have no legal means to challenge it, and they are denied political power or the moral integrity needed for contesting law and power.[44] The foundation of the law, according to Cacho, is predicated upon the exclusion of racialized populations from "its protections and its process of legitimation."[45] As we will see in Chapter 3, biopower does not need this juridical exclusion to kill or let die since eliminating unwanted populations is a positive condition of biopower. Rather, these exclusionary practices serve the murder function of biopower, or what some have called necro-power. Replicants form a nexus with insecurity/exclusion and the city. These complex relationships between the replicant body, insecurity/exclusion, and the city make familiar the ways bodies become legible to the law and neoliberal governmentalities, particularly through the spaces they occupy and the statuses they hold. It is also the case that neoliberalism plays a crucial part in defining and shaping these spaces and statuses.

Replicants, cyborg sex workers, and bio-enhanced gang members (common subject categories in cyberpunk), like poor and racialized populations, are "always already the object and target of the law."[46] They also always already represent a threat of insecurity that the "good" population must be protected from. But *Blade Runner*'s cyborganization of the poor and racialized populations of the present through its characterization of replicants also opens the possibility for viewers to empathize with these populations, and thus, it offers a critical moment for countering the ways criminalization preempts affective (sympathy and empathy) connections with subjects of the law. And, of course, under neoliberalism, the law blames these subjects for their failure to succeed in the game of competition. *Blade Runner* dramatizes the replicants' search and demand for acceptance within the law. Not only are the populations of cyberpunk maximizing their human capital but they are also competing to be defined as human. Replicants must prove they are worthy of the title human.

According to neoliberalism, replicants, cyber-cowboys, cyborg prostitutes, and bio-enhanced gang members (common actors in cyberpunk novels and films) have made a calculated choice to take on an illegal status, and they assume the risks that come with this illegality. Cacho sums up the notion of the entrepreneurial self when she writes that "[n]eoliberal values of private competition, self-esteem, and independence benefit corporations: if everyone is an 'entrepreneur' of him or herself, then individuals cannot be exploited by capital. As 'entrepreneurs' of themselves, individuals exploit themselves and should take 'personal responsibility' for doing so."[47] Corporations are only one among a number of entities that benefit from neoliberal values and the entrepreneurial-self. The notion that individuals exploit themselves and should accept responsibility for this exploitation makes it difficult to assign the origins of exploitation to forms of racism, capitalism, and neoliberalism. However, cyberpunk graphically, perhaps too conveniently, problematizes this misplaced blame by displacing individual responsibility onto advanced forms of capitalism or villainous corporations.

While cyberpunk narrators often attribute failure to a naturalized neoliberal value system (individuals fail because they are lazy, ungraceful, and slow), the protagonists of these stories must at the same time battle the exploitative, violent, and larger-than-life corporate and capitalist forces. As Gilles Deleuze contends, "the corporation constantly presents the brashest rivalry as a healthy form of emulation, an excellent motivational force that opposes individuals against one another and runs through each, dividing each within."[48] Thus, cyberpunk narratives often disrupt the notion that cities are places for individuals to freely compete in the game of competition, a neoliberal game that further romanticizes the city as a space for achieving the "American Dream" (at least in a US context) and often ignores the exploitative and violent influences of neoliberal ideologies and related power forces. Cyberpunk narratives do so by reinserting the role of neoliberal and corporate forces that play a crucial spatial and temporal part in the economic/political/social dynamics of the city. This reinsertion can make it more difficult for states to hide behind neoliberal values by reminding us that the state is actively engaged in producing technologies of security and the conditions that make them necessary.

Denaturalizing the Conditions of Competition

In his book *Punishing the Poor: The Neoliberal Government of Social Insecurity*, Wacquant comments on the naturalization of the market economy. For Wacquant, neoliberals do not turn to nature to justify their "new-style Darwinism." Rather, the market is the setting for the survival of the fittest in the game of competition. However, as Wacquant points out, the market economy is naturalized after it is "depicted under radically dehistoricized trappings which...turn it into a concrete historical realization of the pure and perfect abstractions of the orthodox economic science."[49] I suggest that cyberpunk makes legible, tangible, and spatial the artificial and anthropogenic construction of this naturalized and dehistoricized market economy.

First, cyberpunk historicizes the development of this neoliberal market economy by positioning it after the shift from disciplinary societies to what Deleuze calls societies of control, but what others might call neoliberal biopolitics.[50] This shift to societies of control is, according to Michael Hardt, a shift from modern power and disciplinary power to imperial power and the society of control.[51] Hardt adds that "[i]n its ideal form there is no outside to the world market: the entire globe is its domain...the world market might serve adequately (even though it is not an architecture; it is really an anti-architecture) as the diagram of ...the society of control."[52] Moreover, cyberpunk amplifies to excess the sharing of ideas between "theoretical biology and neoliberal rhetorics of economic growth."[53] Biology and capitalism are both interested in "the limits and possible futures of life on earth."[54] This intertwining of biological fields with economic growth and market economies within cyberpunk makes it easier to see how they are connected as well as biopolitical. For example, in *Neuromancer*, the narrator describes Case's recognition of the economic and spatial logic of the bioeconomies, which are often black markets, and which permeate the streets of Chiba City and Night City. Case indicates that the "burgeoning technologies require outlaw zones, that Night city wasn't there for its inhabitants, but as a deliberately unsupervised playground for technology itself..."[55] And later he states that "[g]enetic materials and hormones trickled down to Ninsei along an intricate ladder of fronts and blinds..."[56]

Along similar lines, the Tyrell Corporation in *Blade Runner* is a thriving bioeconomy that produces genetically engineered human (more than human) slave workers to further the neoliberal biopolitical aim of making certain populations live. These examples illustrate the tight couplings of theoretical biology and economic growth. Cyberpunk's joining of theoretical biology and capitalism within a neoliberal context brings to light their connections in the present, but it also makes conscious their co-emergence in the late-seventies and early-eighties, when many of these cyberpunk novels were written. Thus, cyberpunk can aid in a critique of the "bio-tech revolution"[57] through its amplification of the ways this revolution relocated "economic production at the genetic, microbial, and cellular level, so that life becomes, literally, annexed within capitalist processes of accumulation,"[58] and relied on the highly flexible and continually transformative nature of the capitalist system. The cyberpunk genre can make us sensitive to contemporary capitalism's concerns with the "limits of life on earth and the regeneration of living futures—beyond limits,"[59] since it is often these concerns that unfold within these works.

As I have sought to show above, reading and seeing cyberpunk realigns our scopic plane so that we may recognize "how the earth itself already has submitted to the terraforming of multinational capital formations...," and how "...nature is no longer the vast realm of autonomous, unmanageable, or nonhuman wild activity: in being enmeshed in networks of cyborg scientific rationalization and commercial commodification, nature becomes denatured."[60] As a genre, cyberpunk intensifies informationalization too, that is to say, "the biotronic genesis of transnational capitalism's terraforming of the earth."[61] The earth of cyberpunk is one large built environment. Without these artificial conditions, the neoliberal economic order of cyberpunk could not thrive. There is an overall intensification of the "built environments" of cities. Cyberpunk amplifies the toxic and mutative qualities of the environment. For example, the narrator of *Neuromancer* describes the sky as "the color of television, tuned to a dead channel,"[62] or "the sky was that mean shade of gray. The air had gotten worse; it seemed to have teeth tonight, and half the crowd wore filtration masks."[63] In *Snow Crash*, air pollutants are so thick that they congeal on surfaces if people remain still for too long. In the film *Blade Runner*, the air is thick with smog, the rain is acidic, and the sky is continually dark, at least, up to a certain elevation, suggesting an intensified vision of the consequences of post-industrial pollution.

In addition, the cyberpunk genre amplifies the size of cities, the level of decay, and the number of people in cities. Cities in cyberpunk are integrated with other spaces, which include suburbs and other commercialized zones. As Luke suggests, for cyborgs, "both city and country are architecturally integrated, technologically linked, and institutionally managed to function in corporately administered shelter/diet/energy/dress/labor megamachines."[64] Cyberpunk takes this all-encompassing megamachine to its conceptual/visual limit in that all spaces are anthropogenic constructions. While we can assume that there are other cities in, for example, the future of *Blade Runner*, their possibility falls away in the movie's totalizing vision of Los Angeles. Indeed, the world of cyberpunk is often a combination of endless cities and suburbs. According to Bukatman, cyberpunk illustrates Jameson's suggestion that science fiction "permits the existence of an impossible and totalizing gaze which functions through spatial description...rather than through narrative action."[65] It is partly through this gaze, as I have suggested, that we can make it more difficult for neoliberalism to continue to produce its reality. Cyberpunk brings into focus the very real material consequences of neoliberalism, while reminding us that neoliberalism is a narrative about how we ought to live. Cyberpunk helps make neoliberalism and its worldwide financial networks, hyperobjects as Steven Shaviro suggests,[66] intelligible or conceptually manageable.

Conclusion: The Ambivalent Politics of Cyberpunk

Benjamin Noys and Scott Bukatman both comment on the utopic qualities of cyberpunk. Although Noys' comment addresses Gibson's *Neuromancer*, I think his position on this text is reflective of the cyberpunk genre overall. Noys suggests that the significance of *Neuromancer* and, I would suggest, of all cyberpunk, "lies in the fact that it is poised between anxiety and endorsement, critical distance and immersive jouissance" in its vision and attitude towards cyberspace and other virtual/digital spaces.[67] The drive and valuation of disembodiment come from a privileged position though. Case may reflect a transhumanist desire for and endorsement of bettering the human condition through technological enhancement. However, this transhumanist vision can appear naïve and overly optimistic when we consider the bodies that are incapable and perhaps uninterested in taking bodily transcendence to its limits.

Within the informationalized and cyborganized spaces of cybercities, there is also continual tension between the privileged desire to escape the limitations of the fleshy body and the demand that the corporeal needs and vulnerabilities of some bodies be addressed. In Gibson's *Neuromancer*, Case, the main protagonist, views his body as an encumbrance to his job (as a cyber cowboy, he "jacks" into the matrix, which is a digital reality that is dominated by the flow, transference, and existence of information), since having a body means that he cannot always be in the matrix without dying. For Case, having a body is a weakness and a nuisance. Case's devaluing of his body and embodied experiences points to a romanticization of disembodiment. Disembodiment in cyberpunk is the mind's ability to download, jack-in, or be in a digital, cyber, or virtual reality. It assumes that the mind can exist independent of a biological or organic body, and that existing in a purely informationalized form is better than an embodied, material one. That is, jacking into the matrix, for Case, or being in the Metaverse, for the character Hiro in *Snow Crash*, are ways to transcend bodily limitations (hunger, pain, and death). Thus, devaluing the body and privileging the mind in cyberpunk brings attention to the ways in which neoliberal values and governmentalities are unwilling to recognize how they hide and/or deflect attention away from the systemic causes of economic failure, criminalization, and biological racism. In other words, the values and discourses that permeate the informationalized reality of cyberpunk understand "jacking-in," "plugging-in," or "being-in" a digital reality as a choice, and therefore, as acting through a subject's agency and freedom. Bukatman writes that, while on the surface cyberpunk may have dystopic elements, it can also be seen as a genre that "frequently permits the subject a utopian and kinetic liberation from the very limits of urban existence..."[68] Bukatman adds that much of cyberpunk is interested in the "phenomenologically relevant *other space* of information circulation and control."[69]

Cyber cowboys, in particular, are the ultimate neoliberal subjects since they accept and desire the risk of entering the matrix, of being flexible workers. As flexible workers, cyber cowboys and talented drifters can work wherever they can maximize their competitive edge. The fantasies of disembodiment (the mind/consciousness existing in a purely informational reality without a need for an organic body) distract from the corporeal needs that poor bodies and other devalued bodies must face as they are not in a position to act upon this fantasy. Through valuing disembodiment (the mind) over embodiment (the body), the dominant political and economic order in cyberpunk can reinvigorate ignoring the bodies of the classes they exploit and let die. Once again, if a person experiences bodily suffering, it is the result of their inability to compete in a system that valorizes digital ontologies. However, cyberpunk is also not consistent with its valorization of digital ontologies/disembodiment/the possessive individual. That is, it critiques the environmental consequences of these digital ontologies, and at the same time, it glorifies these ontologies. Cyberpunk is also often unaware of the fact that it valorizes the kind of political economy and subjectivity that perpetuate neoliberal governmentalities and produce the environmental death it critiques. For example, *Neuromancer*'s Case does not question the neoliberal logic of Night City's black market bioeconomies, where insecurity is necessary. Instead, he sees a kind of natural logic to these spaces. It is not entirely clear whether Gibson is celebrating instrumental reason, or if he is showing the physical and material conditions that neoliberal individuals must face in denatured/post-industrialized environments. Thus, it is perhaps almost by accident that cyberpunk brings to attention the everyday suffering that permeates its streets as its digital dramas unfold.

While cyberpunk is at times clueless about the material consequences of the political economy and instrumental rationality within which neoliberalism and denatured environments thrive, I argue that this naivety/cluelessness, and, perhaps, even at times, a willful ignorance, is what makes cyberpunk an especially (ab)useful genre for producing epistemological interferences. Borrowing from Timothy Morton's work on the "hyperobject," Steven Shaviro defines hyperobjects as "…phenomena that actually exist but that 'stretch our ideas of time and space, since they far outlast most human time scales, or they're massively distributed in terrestrial space and so are unavailable to immediate experience.'"[70] Shaviro argues that it is through science fiction that we can try to make these hyperobjects intelligible/recognizable. The vast financial networks of cyberpunk can be seen as hyperobjects. But, as a genre, cyberpunk also works at making these hyperobjects available to immediate experience. Cyberpunk may at times uncritically valorize instrumental rationalities and privilege a cartesian mind/body dualism as its protagonists strive towards digital being. But, in doing so, it also brings into focus the struggles that neoliberal subjects face in recognizing or dealing with the very conditions (and information abstractions) that make it difficult to see how neoliberal economics works in any concrete terms, at least beyond the constant crises of neoliberalism. Cyberpunk highlights the struggles of individual subjects to compete and make choices, as they are encouraged to do by neoliberal values, and at the same time, to make sense of and integrate with hyperobjects (the financial networks/digital networks). As aesthetic subjects, Case, Hiro, and Takeshi Kovacs make familiar the abstract forces (information flows, financial networks, and other economic processes) at play within the spaces produced and ordered by neoliberalism and its governmentalities.

Cyberpunk does not always offer a consistent sophisticated critique of neoliberalism and its abstractions. Still, its aesthetic subjects "articulate and mobilize thinking" that is useful for producing discomfort within our current conditions. That is, these aesthetic subjects think like neoliberal subjects (they assume all life risks, they understand freedom as unhindered choice, and they instrumentalize self-cultivation to maximize human capital), and through their thinking, they make familiar some of the values and discourses they encounter and practice in the spaces of competition. Chapter 2 examines a concrete example of the cyberpunk neoliberal subject: the self-monitoring cyborg. It focuses on the self-monitoring cyborg in order to highlight the ways in which neoliberalism, its discourses, and its governmentalities capitalize on notions of self-cultivation to hide the fact that these individuals are fully subsumed under capitalism.

Notes

1. Michael Shapiro, *The Time and the City: Politics, philosophy and genre* (New York: Routledge, 2010).
2. Timothy Luke. "The Co-Existence of Cyborgs, Humachines and Environments in Postmodernity: Getting Over the End of Nature," in *The Cybercities Reader* (New York: Routledge, 2004), 109.
3. Timothy Luke, "Liberal Society and Cyborg Subjectivity: The Politics of Environments, Bodies, and Nature," *Alternatives: Global, Local, Political*, 21, no. 1 (1996): 3.
4. Ibid., 3.
5. According to Luke, nature "...in being enmeshed in networks of cyborg scientific rationalization and commercial commodification, nature becomes denatured." Furthermore, denature is "nature, deformed and reformed simultaneously by anthropogenic transformations...Whatever nature once was cannot be regained, because it existed as a set of forces, settings, or conditions when the human influences upon planetary ecologies were very low impact" See Luke "At the End of Nature: Cyborgs, 'Humachines,' and Environments of Postmodernity," 109. In the context of Luke's cyborg theory, "dehumanity," "dehumanization," and "dehumanized," do not refer to the liberal anxiety about the loss of the category of human and all of its corresponding rights and political power. Instead, for Luke, "dehumanized" beings "inhabit the modernized global ecologies of mechanized, polluted, bioengineered denature as fragments and fusions of the machinic systems that define today's environments, bodies, and politics" See Luke, "Liberal Society and Cyborg Subjectivity," 7. Luke accepts the dehumanized condition as a constant, rather than as a threat to liberal human status.
6. Ibid., 3.
7. Tim Luke, "At the End of Nature: Cyborgs, 'Humachines,' and Environments of Postmodernity," in *Environment and Planning A* 29 (1997): 1368.
8. Fredric Jameson, "Progress versus Utopia, or, Can We Imagine the Future?" *Archaeologies of the Future: The Desire Called Utopia and Other Science Fictions* (London: Verso, 2005), 286.
9. Scott Bukatman, *Terminal Identity: The Virtual Subject in Postmodern Science Fiction* (Durham: Duke University Press, 1993), 146.
10. Ibid., 86.
11. Melinda Cooper, *Life as Surplus: Biotechnology and Capitalism in the Neoliberal Era* (Seattle: University of Washington Press), 20.
12. In *No Speed Limit*, Steven Shaviro writes that accelerationism is "the argument that the only way out is the way through. In order to overcome globalized neoliberal capitalism, we need to drain it to the dregs, push it to its most extreme point, follow it into its furthest and strangest consequences." 2. More on accelerationism in Chapter 5.
13. Steven Shaviro, *No Speed Limit: Three Essays on Accelerationism* (Minneapolis: The University of Minnesota Press, 2015), 2.
14. Ibid., 7.
15. Michel Foucault, *The Birth of Biopolitics: Lectures at the Collège de France, 1978–79*, trans. Graham Burchell (New York: Palgrave MacMillan, 2008), 120.
16. Loïc Wacquant, "Social Insecurity and the Punitive Upsurge," in *Punishing the Poor: The Neoliberal Government of Social Insecurity* (Durham: Duke University Press, 2009), 8.
17. Foucault, "*The Birth of Biopolitics*," 243.
18. Ibid., 242.

19. Wacquant, *Punishing the Poor*, 8.
20. Ibid., 246.
21. Ibid., 246.
22. William Gibson, *Neuromancer* (New York: Ace, 1984), 7.
23. Lisa Marie Cacho, *Social Death: Racialized Rightlessness and the Criminalization of the Unprotected* (New York: New York University Press, 2012), 20.
24. Gibson, *Neuromancer*, 7.
25. Thomas Lemke, "The Risks of Security: Liberalism, Biopolitics, and Fear," in *The Government of Life: Foucault, Biopolitics, and Neoliberalism*, eds. Vanessa Lemm and Miguel Vatter (New York: Fordham University Press, 2014), 68.
26. Aihwa Ong asserts that neoliberal governmentality finds its origins in Foucault's biopower, and that neoliberalism is "merely the most recent development" of biopolitics. See Aihwa Ong, *Neoliberalism as Exception: Mutations in Citizenship and Sovereignty* (Durham: Duke University Press, 2006), 13.
27. Lemke, "The Risks of Security," 68.
28. Ibid., 65.
29. Neal Stephenson, *Snow Crash* (New York: Bantam, 1992), 6.
30. Mike Davis, "Fortress L.A.," in *City of Quartz* (New York: Vintage, 1992), 250.
31. I return to bio-enhancement in Chapter 5. Here, I examine the ways in which neoliberalism has intensified, similar to the ways cyberpunk amplifies bio-enhancements, the individual, particularly in the form of a bio-hacker. Intensification is one way in which neoliberalism furthers its reality.
32. Benjamin Noys, "Cyberpunk Phuturism," in *Malign Velocities: Acceleration and Capitalism* (Washington: Zero Books, 2014), 52.
33. The non-radical reality of cyborg formations points to the ways in which cyberpunk's cyborganization is unlike Haraway's cyborg. The politics of cyberpunk, as I suggest later in this chapter, are largely ambivalent about its radical possibilities. Cyborgs in cyberpunk still exist within a biopolitical system. Cyborganization is largely about the ways in which individuals make themselves more competitive as neoliberal subjects.
34. I will further develop the anchoring of this study in the context of biopolitics in Chapter 3.
35. Richard K. Morgan, *Altered Carbon* (New York: Del Rey, 2002). The term "sleeve" in *Altered Carbon* refers to the way individual minds can wear bodies like a piece of clothing.
36. I am primarily concerned with the original *Blade Runner* in its Final Cut version (2007).
37. Sharon Holland, *Raising the Dead: Readings of Death and (Black) Subjectivity* (Durham: Duke University Press, 2000), 15.
38. Cacho, "Social Death," 7.
39. Ibid., 7.
40. On terraforming, see Luke, "Co-Existence of Cyborgs," 107.
41. In *"Society Must Be Defended,"* Foucault defines racism as "primarily a way of introducing a break into the domain of life that is under power's control: the break between what must live and what must die." See Michel Foucault, *"Society Must Be Defended": Lectures at the Collège de France, 1975–76*, translated by David Macey (New York: Picador, 1997), 254.
42. Cacho, *Social Death*, 6.
43. Ibid., 6.
44. Ibid., 6.

45. Ibid., 5.
46. Ibid., 5.
47. Ibid., 19.
48. Gilles Deleuze, "Postscript on the Societies of Control," *October* 59 (1992): 5.
49. Wacquant, *Punishing the Poor*, 6.
50. See Aihwa Ong, *Neoliberalism as Exception: Mutations in Citizenship and Sovereignty* (Durham: Duke University Press, 2006); Melinda Cooper, *Life as Surplus: Biotechnology and Capitalism in the Neoliberal Era* (Seattle: University of Washington Press, 2008); T*he Government of Life: Foucault, Biopolitics, and Neoliberalism*, edited by Vanessa Lemm and Miguel Vatter (New York: Fordham University Press, 2014), among others.
51. Michael Hardt, "The Global Society of Control," *Discourse* 20, no. 3 (1998): 143.
52. Ibid., 143.
53. Cooper, *Life as Surplus*, 20.
54. Ibid., 20.
55. Gibson, *Neuromancer*, 11.
56. Ibid., 11.
57. Cooper, *Life as Surplus*, 19.
58. Ibid., 19.
59. Ibid., 20.
60. Luke, "The Co-Existence," 108
61. Ibid., 108.
62. William Gibson. *Neuromancer*, 3.
63. Ibid., 15.
64. Luke, "Liberal Society," 12.
65. Scott Bukatman, "The Cybernetic (City) State: Terminal Space Becomes Phenomenal," *Journal of the Fantastic in the Arts 2*, no. 2 (1989): 52.
66. Steven Shaviro, *No Speed Limit: Three Essays on Accelerationism* (Minneapolis: University of Minnesota Press, 2015), 9.
67. Noys, *Malign Velocities*, 57.
68. Bukatman, *"Terminal Identity,"* 145.
69. Ibid., 94.
70. Shaviro, as cited in *No Speed Limit*, 9.

Bibliography

Bukatman, Scott. *Terminal Identity: The Virtual Subject in Postmodern Science Fiction*. Durham: Duke University Press, 1993.

—., Scott. "The Cybernetic (City) State: Terminal Space Becomes Phenomenal." *Journal of the Fantastic in the Arts 2*, no. 2 (1989): 42–63.

Cacho, Lisa Mari. *Social Death: Racialized Rightlessness and the Criminalization of the Unprotected*, New York: New York University Press, 2012.

Cooper, Melinda. "Life beyond the Limits," in *Life as Surplus: Biotechnology and Capitalism in the Neoliberal Era*, Seattle: University of Washington Press, 2008: 15–50.

Davis, Mike. "Fortress L.A.," in *City of Quartz*, New York: Vintage, 1992: 221–264.

Deleuze, Gilles. "Postscript on the Societies of Control," *October* 59 (1992).

Foucault, Michel. *The History of Sexuality: An Introduction, Volume I*, New York: Vintage Books, 1978.

—., *The Birth of Biopolitics: Lectures at the Collège de France, 1978–79*, Translated by Graham Burchell, New York: Palgrave MacMillan, 2008.

Gibson, William. *Neuromancer*, New York: Ace, 1984.

Hardt, Michael. "The Global Society of Control," *Discourse* 20, no. 3 (1998).

Holland, Sharon. *Raising the Dead: Readings of Death and (Black) Subjectivity*, Durham: Duke University Press, 2000.

Jameson, Fredric. "Progress versus Utopia, or, Can We Imagine the Future?" in *Archaeologies of the Future: The Desire Called Utopia and Other Science Fictions*, London: Verso, 2005: 281–295.

Lemke, Thomas. "The Risks of Security: Liberalism, Biopolitics, and Fear," in *The Government of Life: Foucault, Biopolitics, and Neoliberalism*, edited by Vanessa Lemm and Miguel Vatter, New York: Fordham University Press, 2014.

Luke, Timothy. "Liberal Society and Cyborg Subjectivity: The Politics of Environments, Bodies, and Nature," *Alternatives: Global, Local, Political* 21, no. 1 (1996).

—., "At the End of Nature: Cyborgs, 'Humachines,' and Environments of Postmodernity," *Environment and Planning A*, 29 (1997).

—., "The Co-Existence of Cyborgs, Humachines and Environments in Postmodernity: Getting Over the End of Nature," in *The CyberCities Reader*, edited by Graham, Stephen, New York: Routledge, 2004.

Morgan, Richard K. *Altered Carbon*. New York: Del Rey, 2002.

Noys, Benjamin. "Cyberpunk Phuturism," in *Malign Velocities: Acceleration and Capitalism*, Washington: Zero Books, 2014: 49–62

Ong, Aihwa. *Neoliberalism as Exception: Mutations in Citizenship and Sovereignty*, Durham: Duke University Press, 2006.

Shapiro, Michael. *The Time of the City: Politics, Philosophy and Genre*, New York: Routledge, 2010.

Shaviro, Steven. *No Speed Limit: Three Essays on Accelerationism*, Minneapolis: University of Minnesota Press, 2015.

Stephenson, Neal. *Snow Crash*, New York: Bantam, 1992.

Wacquant, Loïc. "Social Insecurity and the Punitive Upsurge," in *Punishing the Poor: The Neoliberal Government of Social Insecurity*, Durham: Duke University Press, 2009: 1–40.

2 Self-Monitoring as Instrumentalized Self-Cultivation

Introduction

Jonathan Crary argues in his book *24/7: Late Capitalism and the Ends of Sleep* that, under neoliberalism, individuals model themselves, often with the help of new information technologies, based on the logic of the market, information, and machinic systems. While Crary offers a compelling critique of the ways in which neoliberal labor time has colonized all facets of life, his use of the word "modeling" here suggests an intentionality and level of agency on the part of the individuals doing the modeling that I would suggest is often absent. And, his claim that, with the exception of sleep, all time has been subsumed under work/market time does not go far enough. Crary's use of "personal and social identity" does not capture the subjectivities that have developed out of the conditions of competition and according to the logic of intensity that are part of neoliberalism today. I would like to take Crary's argument further to, one, suggest that "identity" is not enough to account for what exactly has been "reorganized to conform to the uninterrupted operation of markets, information networks, and other systems,"[1] and, two, to argue that Crary's 24/7 work time does include sleep.

The new information technologies Crary references—I will specifically address self-monitoring devices/practices and the subjectivities they arouse—encourage individuals to monitor and manage their sleep. Sleep is not a repose from capital or labor time, as Crary suggests. While I am not primarily concerned with sleep in this chapter, I do think the extension of work time, the economic, and the maximization of human capital into sleep, points to an intensification of the entrepreneur of the self and a logic of intensity that Crary and perhaps other Marxist inspired theorists miss. Sleep must now be monitored to ensure that individuals are sleeping the appropriate amount, in the appropriate way, at the appropriate time, in order to maximize optimal performance/human capital and remain competitive. Sleep time is now work time. In other words, when individuals sleep, they are working. They are working at selling themselves. Sleep has been subsumed under a form of "self-cultivation" that is itself no longer antithetical to late capitalism or neoliberalism. Populations of individuals have instrumentalized self-cultivation, a notion and practice that is often still imagined as something that stands outside of work time and capital time, in order to maximize their human capital and live intensely.

I suggested in the Introduction that, out of the interplay between certain neoliberal discourses (in the case of this chapter, neoliberal health and self-monitoring/quantification) and certain nodes of power-knowledge, emerges a particular thinking subject that understands itself in economic terms, as a homo-economicus, or an entrepreneur of the self. This thinking subject is more than a personal or social identity. Neoliberal subjects are not consistently conscious of the fact that they are subjectivizing themselves based on the ways in which neoliberal governmentality spreads its subjectivity. If, as Thomas Lemke argues, neoliberalism "endeavors to create a social reality that it suggests already exists,"[2] then individuals may not always be aware of what exactly is "molding," "reorganizing," "regulating," or "producing" the possibility of their subjectivity or the fact that they are actively molding themselves based on the logic of the market. Instead, populations of individuals may see the "uninterrupted operation of markets, information networks, and other systems"[3] and the economization of the social as the way it has always been or just the way it is today. Even if there are new technologies that enable individuals to engage in seemingly new practices, such as quantifying the self and biohacking, these practices are predicated on a political, social, and economic project/reality/condition, namely neoliberalism, that has been around for over 35 years.

In Chapter 1, I indicated that neoliberalism has capitalized on the continued existence of liberalism and, I would add, on notions of self-cultivation because liberalism and self-cultivation, which are often contradictory and antithetical to neoliberalism (e.g., self-cultivation stands outside an economic logic and liberalism accepts that competition is natural), hide the ways some governmentalities intensify processes of competition and displace responsibility onto individuals. This chapter considers concrete examples of the cyberpunk subjectivities I introduced in Chapter 1 (the cyber cowboy and the cyborg in particular). Similar to Chapter 1, then, this chapter continues the project of making familiar the ways in which neoliberalism maintains its reality, since as I suggested above individuals are not always aware of the productive forces of neoliberalism. The cyberpunk subjectivities of cyberpunk science fiction are neoliberal subjects, and like the cyber cowboy or the cyborg of cyberpunk, through their daily practices, the self-monitoring cyborg makes neoliberalism possible by intensifying notions of the individual and by practicing "self-cultivation." Thus, a goal of this chapter is to bring attention to some of the ways neoliberal subjects do neoliberalism by focusing on several everyday practices, such as exercise, self-monitoring, sleep, calorie counting, etc. Notions of self-cultivation are cyberpunk romantic stories we tell ourselves to make maximizing human capital a more noble endeavor.

I focus on the self-monitoring cyborg in order to highlight the ways in which neoliberal governmentalities, including nodes of power-knowledge (e.g., the co-productive relationship between the tech industry and neoliberal health discourses), capitalize on notions of self-cultivation/self-care that on one level hide the fact that, as Steven Shaviro puts it,

> ...labor, subjectivity, and social life are no longer "outside" capital and antagonistic to it. Rather, they are immediately produced as parts of it. They cannot resist the depredations of capital, because they are themselves already functions of capital. This is what leads us to speak of such things as "social capital," "cultural capital," and "human capital": as if our knowledge, our abilities, our beliefs, and our desires had only instrumental value and needed to be invested... [e]verything without exception is subordinated to an economic logic, an economic rationality.[4]

And, on another level, capitalizing on notions of self-cultivation hides the fact that neoliberalism produces, nurtures, and benefits from a particular kind of subject, in this case the self-monitoring cyborg, that desires, valorizes, and normalizes the assumption of risk and the responsibility for his/her/its choices, in order to successfully compete. In this context, competing equates a life worth living with living life. If the neoliberal subject does not compete, then it is not living. The neoliberal subjects that fail to compete endure a quasi-life. They are not quite living, and they are not quite dead; although, they are definitely closer to a living death than more competitive populations of individuals. I am reminded of the ways in which Case of Gibson's novel *Neuromancer* lives by a terminal overdrive, always on the edge of burn out. Case must "heed the demands of an intricate protocol,"[5] a protocol that is informed by competition.

I am not suggesting that the valorization of competition is acceptable as long as there is a social safety net. Rather, in this chapter, I will highlight at least one way in which the neoliberal subject kills itself and tries to remain relevant to/and intelligible with regards to a market logic by happily eliminating variations of itself in order to maximize human capital and endure intense physical and emotional stress to compete. And, as I will show in Chapter 5, the self-monitoring cyborg is only the beginning in the game of intensification, since in order to compete even processes like self-monitoring and notions of self-cultivation do not go far enough; they must be intensified. Chapter 5 shows that the biohacker and self-mastery are ways in which we have intensified, or amplified, life to avoid burn out.

Neoliberal subjects accept risk and are always at risk. These neoliberal subjects, including the self-monitoring cyborg, are always at risk of burnout because they are subjectivized to live life intensely. As Jeffrey Nealon claims, neoliberalism functions, in part, through a logic of intensity.[6] Robin James takes this logic of intensity further in her article "Loving the Alien." She argues that "[r]iding the crest of burnout is associated with privilege. Hegemony reproduces itself by distributing resources to privileged groups; thus, privileged people get to lead the most intense lives, lives of maximized (individual and social) investment and maximized return. Experientially, privilege means being so busy, overcommitted, and invested in your life that you're always at risk of hitting the point of diminishing returns."[7] Even less privileged individuals are subjectivized to live life intensely (e.g., fitness tracking devices incite middle to lower-income individuals to intensify their assumption of risk and responsibility for their health choices in order to maximize their human capital). But they do not have the same level of resources as privileged groups, and thus, are more likely to hit the point of diminished returns. Self-monitoring encourages individuals to be highly invested in their life, to devote 24/7 to maximizing their human and social capital, and to continually live life at the edge of burn out. Most of us just do not have the resources to avoid burn out, to avoid a life of struggling to endure while living life by this logic of intensity. Populations of individuals are, as Huawaei puts it,[8] incited to "pursue their dreams," to be "people who are visionary challengers," to be "people who are proactive with a strong belief in striving to achieve their goals," to "[create] extraordinary experiences for people everywhere."[9]

I understand self-cultivation as practices that aim towards caring for the self as an end in itself, not necessarily as a means to compete and sell oneself on the market. Self-cultivation understood as caring for the self may include some of the same practices as self-monitoring cyborgs, including exercise, changing eating habits, and taking supplements. But the difference is that self-cultivation is not a function of capital. There may be no particular end game to acts of cultivating the self. Perhaps, cultivating/improving/flourishing are enough. In other words, exercise does not have to be a means to improve or flourish, but rather is itself improvement or flourishing. Self-cultivation is not necessarily dependent upon its instrumental value. Self-cultivation is something to "enjoy entirely for its own sake."[10] What the self-monitoring cyborg does by monitoring its sleep, tracking its exercise and calorie intake, regulating and policing other individuals' monitoring/tracking/quantifying habits, exercising while collecting data, is not self-cultivation. Instead, its practices are functions of capital and neoliberalism under the guise of self-cultivation. By self-monitoring cyborg, I mean simply the individuals who join (sometimes temporarily and sometimes permanently) fitness/health tracking devices and applications to their body to "self-cultivate."

I am not using cyborg to refer to a transgressive, radical, or liberatory ontology or subject here. In fact, it is just the opposite as there is nothing liberatory or transgressive about this modality of self-monitoring cyborganization.

The reality of the cyborg suggests that it does not exist outside of neoliberalism and its discourses/governmentalities. Rather, modern cyborg ontologies, like the self-monitoring cyborg, often function as strategies of governmentality, in this case neoliberal governmentality. Wearing a fitness tracker enables and incites the user to practice neoliberal subjectivity. In short, neoliberal governmentality utilizes the normalizing medical discourses that lend fitness trackers authority over bodies to eliminate unwanted bodies, including obese bodies and sedentary bodies, since both of these bodies are incapable of effectively competing and selling themselves as capital in a market economy. Rather than offering the cyborg as a transgressive and potentially liberatory ontology, this chapter will suggest, instead, that the self-monitoring cyborg is both an apparatus of biopower and a strategy of neoliberal governmentality in that it re-inscribes bodies within neoliberal discourses. Haraway writes that "the cyborg is not subject to Foucault's biopolitics."[11] However, I argue that as a strategy of neoliberal governmentality the fitness tracking cyborg is a technique of biopower. That is, it offers a means for governmentality to govern life by eliminating unwanted bodies, and it allows subjects to govern themselves by furthering the quotidian practices of neoliberal subjectivity. Quotidian practices of a particular kind of neoliberal subject, in this case the self-monitoring cyborg (I will consider the biohacker in the final chapter), are aimed towards maximization and intensification. The life and living that is proper to neoliberal biopower is governed by the logic of competition. Populations of individuals govern themselves, in part, through maximization and intensification in order to meet the demands of competition. Living intensely and maximizing in order to compete is indicative of neoliberal biopower. I will begin by offering a brief history of fitness trackers and reviewing popular techno discussions and recent studies of fitness trackers to show how the fitness tracking cyborg is an apparatus of neoliberal biopower. Specifically, this section will consider the neoliberal discourses and medical discourses on health and exercise that normalize/maximize bodies. Then I will offer an analysis of online Fitbit communities, including discussion boards and community groups, as concrete examples of how a fitness tracker is again a strategy of neoliberal biopower, and I will theorize how the self-monitoring cyborg lives intensely. I draw from the language of discussion boards and community groups in order to highlight one way neoliberalism as a systemic reality is perpetuated. These discussion boards and community groups do neoliberalism as they reproduce the cyberpunk and romantic narratives I have explored thus far, including stories about self-cultivation or maximization, individualism, competition, resilience, and living the most intense lives.

An Analysis of Popular Techno Discourses

Over the last decade, a new technology has emerged in the health and exercise market in response to a demand for self-regulatory and motivational mechanisms: the fitness tracker. A fitness tracker is a form of wearable technology that users can wear on their bodies partnered with application software in order to monitor their physiological progress as they move or exercise throughout their day. In their nascent form, fitness trackers were basic pedometers that users could attach to their clothing. These pedometers counted steps and the number of miles for individual users. Pedometers have transformed into more sophisticated wearable technologies that allow users to record, view, and share their information with the public. In their current form, fitness trackers can be worn on the wrist and around the waist in order to measure steps taken, miles walked or jogged, calories burned and consumed, hours slept successfully, stairs ascended, etc. Wearers can also monitor their daily water and food intake, along with their weight and heart rate in an application that functions on their smart devices and laptop computers.

In 2007, Gary Wolf and Kevin Kelly coined the phrase "quantified self" to refer to the growing capabilities humans had to monitor and collect data on themselves. In the past, researchers and consumer analysts collected and provided data to the public. However, after the production of wearable technologies and self-monitoring software, individuals could gain access to real-time information about their athletic and day-to-day physiological performance without the need for researchers or analysts to produce the data for them. In considering what fueled this still growing trend journalists often turn to an increase in data-tracking devices, suggesting the less advanced forms of data-recording processes were already there (journaling, diaries, and spreadsheets) just waiting for a more efficient form of monitoring to appear on the market.

Gary Wolf, a well-known writer for *Wired*, wrote in 2009, before the development of the most advanced fitness trackers, that his colleagues were,

> finding clever ways to extract streams of numbers from ordinary human activities. A new culture of personal data was taking shape. The immediate cause of this trend was obvious: New tools had made self-tracking easier. In the past, the methods of quantitative assessment were laborious and arcane.[12]

Wolf does not offer a reason for why ordinary people would want to self-track. I want to suggest that this drive for self-tracking is motivated by a political economy of normalized bodies interested in the analysis of everyday activities. This economy of self-monitoring bodies is, as Foucault suggests in "*The Birth of Biopolitics*", "the extension of economic analysis into a previously unexplored domain, and second, on the basis of this, the possibility of giving a strictly economic interpretation of a whole domain previously thought to be non-economic..."[13] The production of self-monitoring technologies answers the neoliberal demand for a way of making available information about human behavior that can be used to further competition.

Other tech-writers have looked to human nature as the cause of the rise of self-monitoring technologies. Ryan Holmes, a self-professed "serial entrepreneur" and popular tech-journalist, naturalizes self-monitoring in his suggestion that these new technologies appeal to human nature. He writes, "I think on the most basic level the movement taps into deep-seated human impulses—the quest for self-knowledge and self-improvement, with a touch of narcissism thrown in. By carefully tracking the metrics that define our physical lives, it's possible to identify trends, make adjustments and, in theory, get a bit better at whatever we're doing—from running 10Ks to trying to get a good night's sleep."[14] Holmes suggests that users are motivated by natural drives for self-monitoring. Normal and good people should be concerned with knowing and improving themselves, but they often do both towards the goal of bettering their competitive edge, not necessarily for the intrinsic value of self-knowledge.

Wolf and Holmes echo the thoughts of a number of tech-journalists; often they take for granted the science behind these fitness tracking devices, assuming it is only natural that individuals would want to take control of their health. Current self-monitoring devices like fitness trackers allow users to collect and share a vast amount of what would have been considered private information, perhaps only shared with a general practitioner, with the public via applications on their smart phone, social media, and in online communities. For Foucault, it is "no longer the analysis of the historical logic of processes; it is the analysis of the internal rationality, the strategic programming of individuals' activity..."[15] Wolf and Holmes's analysis of self-monitoring technology cannot see the biopolitical agenda of the medical discourses behind the "science" of these devices. Rather than tracing the appeal of this technology to convenience or human nature, we can look to a "self-imposed" social problem with improving human capital. The case can be made that the fitness tracker, as a type of self-monitoring device, works along the side of medical discourses to strategically program individuals' activities for self-improvement. Foucault contends that, "...as soon as society poses itself the problem of improvement of its human capital in general, it is inevitable that the problem of the control, screening, and improvement of the human capital of individuals, as a function of union and consequent reproduction, will become actual, or at any rate, called for."[16]

Neoliberalism has posed the problem of improvement in response to its abnormal and lazy bodies. The answer to this problem is the process of normalizing these bodies so that they can be productive and competitive. Hence, fitness trackers enable individuals to increase their entrepreneurial potentialities by multiplying their self-monitoring techniques. Users can turn a critical eye on their bodies, view the information their trackers report, and eliminate the unwanted parts through exercise, diet, or surgery.

After using a variety of fitness trackers in order to discover their effectiveness, tech-journalist Albert Sun wrote: "[t]hese days, I've become keenly aware of how active I am and how active I need to be in order to feel healthy and energized. I don't need a monitor anymore. I'm tracking me."[17] To a degree, fitness trackers act as a set of training wills for users to develop, or in some cases heighten, self-monitoring and self-improving habits. In some cases, once users take on the monitoring practices of the fitness tracking software, they continue to regulate their bodies as fitness tracking machines. If individuals do not have the benefit of "good genes," they can enhance their bodies by working hard and increasing their efficiency. They can internalize neoliberal values, such as responsibility, risk-assumption, intensity, and resilience. Fitness tracking devices are a form of anatomo-politics and biopolitics. Foucault examines both in "The Right of Death and Power Over Life"; he writes that one of these forms "centered on the body as a machine: its disciplining, the optimization of its capabilities, the extortion of its forces, the parallel increase of its usefulness and its docility, its integration into systems of efficient and economic controls, all this was ensured by the procedures of power that characterized the disciplines..."[18] Human users transform their bodies into more efficient machines when they join fitness trackers to their body. Through the use of fitness trackers, individuals discipline their bodies by monitoring their daily physiological information, and then by changing their behavior in order to produce better bodies.

The medical science behind fitness trackers helps regulate different populations of individuals. This regulatory power over life is, once again, what Foucault calls biopower. Biopower focuses on "the species body, the body imbued with the mechanics of life and serving as the basis of the biological processes: propagation, births and mortality, the level of health, life expectancy and longevity, with all the conditions that can cause these to vary."[19] In 2011, *The Journal of Medicine and Science in Sports and Exercise* released their recommendations for increasing health, life expectancy, and longevity through exercise and physical fitness. According to the report, adults should exercise for thirty minutes, five days a week, including cardiorespiratory, muscular, flexibility, and neuromotor activities. This report makes it clear that healthy or, at least, healthy defined this way, should be the norm, and it advises health and exercise experts on ways to guide and motivate adults towards increased levels of activity.[20] It prescribes behavioral changes for the masses in order to make the population healthier overall. Biopower increases the authority of medical institutions as they observe, diagnose, treat, and prescribe the public. Out of the shift towards biopolitics, Foucault writes that "[a] 'medico-administrative' knowledge begins to develop concerning society, its health and sickness, its conditions of life, housing, and habits...And there likewise constituted a politico-medical hold on a population hedged in by a whole series of prescriptions relating not only to disease but to general forms of existence and behavior..."[21] The prescriptive report released by *The Journal of Medicine and Science in Sports and Exercise* answers the need for knowledge about bodies, and it increases the hold medicine and health have over the population. Fitness trackers, self-monitoring technologies, and medical reports work together as techniques of biopower; they function as a normalizing science where knowledge and power intensify each other.

Furthermore, the report appeals to the belief that, if individuals can just take responsibility for their bodies, they can overcome their genes, prolong their lives, and maximize their human capital. Foucault's "The Politics of Health in the Eighteenth Century" highlights the role health plays in political power, suggesting that "the emergence of the health and physical well-being of the population in general became one of the essential objectives of political power"[22] during the 18th century in Europe. Health became "at once the duty of each and the objective of all."[23] From this point on, the public takes for granted the rationality behind using self-monitoring technology as though it is only "natural" that individuals would want to take control of bettering their bodies. Thus, using fitness trackers is both individualizing since it enables users to discipline their bodies and practice neoliberal subjectivity and "massifying" since it brings humans into the calculable matrix of a population.

Donna Haraway suggests that "[m]odern medicine is also full of cyborgs, of couplings between organism and machine, each conceived as coded devices, in an intimacy and with a power that was not generated in the history of sexuality."[24] Medical discourses and neoliberalism work together in a tight "cluster of relations" to produce self-monitoring cyborgs that function as technologies of power. An individual does not necessarily transgress boundaries by connecting a fitness tracker to their body. Even in more extreme forms, the cyborg may still be a way to practice forms of being that perpetuate the metaphysics of dominant discourses and networks of power. Neoliberalism encourages subjects to alter their bodies if by doing so it increases the value of their human capital.

As citizens continue to regulate and cultivate their bodies, neoliberal governmentality's role in governing becomes hands off. Fit bodies are ideal because they do not require as much governing and because they are cheaper. Individuals think they are bettering themselves when they use fitness trackers on their own terms, that losing weight and exercise is a sign of their character and agency. But while they think and do so, they are making themselves better at self-subjugation. Some users immediately derive pleasure from regulating, cultivating, and monitoring their bodies, and others experience pleasure over time.

The dominant neoliberal rhetoric surrounding fitness trackers portrays these devices as empowering, uplifting, motivational, essential, and life-changing. Corporate mission statements reveal an underlying desire to make the body more competitive in its ability to sell itself. These statements are in line with a neoliberal mode of living whereby the human is an entrepreneur of the self. For example, the Fitbit "Who We Are" statement reads: "We're a passionate team dedicated to health and fitness, who are building products that help transform people's lives. While health can be serious business, we feel it doesn't have to be. We believe you're more likely to reach your goals if you're encouraged to have fun, smile, and feel empowered along the way."[25] The suggestion here is that people want or need to be empowered and transformed, and that the Fitbit can do both. Fitbit users should also have fun and smile as they use their fitness tracking device. They are tapping into a health apparatus that leads people to believe they can and should have the power to transform their bodies. This statement shows that the makers of Fitbit are working to normalize a lifestyle surrounding the use of the Fitbit. The Fitbit website includes pictures of the Fitbit team; all of the team members are smiling, which is consistent with the happy, having fun attitude that goes with Fitbit rhetoric.

The Fitbit mission is: "To empower and inspire you to live a healthier, more active life. We design products and experiences that fit seamlessly into your life so you can achieve your health and fitness goals, whatever they may be."[26] The Fitbit description statements promote a way of being that is in line with neoliberal narratives about motivation and hard work. Fitbit, Inc., claims that the Fitbit empowers its users. Yet it also regulates users through the expectations that are governed by health and medical fields about and for users. Thus, while it may seem as though Fitbit empowers users by providing them access to useful information that contributes to their overall health, I would suggest instead that the Fitbit simultaneously compels users to discipline their bodies and enables neoliberal governmentality to regulate populations. Foucault makes it clear in *The History of Sexuality: Volume I* that "...[p]ower is tolerable only on condition that it mask a substantial part of itself. Its success is proportioned to its ability to hide its own mechanisms."[27] The Fitbit mission statements hide the underlying force relations that are at work on the body by couching them in the language of empowerment and transformation.

Similar to the makers of Fitbit, the Basis fitness tracker offers an entire page devoted to justifying the use of its fitness trackers. Basis offers a series of answers to the question "Why use fitness trackers?" as part of their mission statement. They begin with the following: "[w]hen it comes to getting fitter and healthier, knowledge is power. And a new breed of wrist-worn gadgets can tell you more about yourself than even you ever knew."[28] What kind of power is Basis referring to? Perhaps it is individual power over the body; the power to discipline our bodies based on commonly accepted bodily norms. This statement implies that people should want to know as much about their bodies as possible so that they can make sure that their body is normal and desired, rather than abnormal and unwanted. Here again, knowledge and power appear together, and they intensify each other. The will to knowledge about one's body enhances the technologies and techniques power has over the body. Individuals must seek out the abnormalities and ugliness of health that are part of self-disciplining their bodies. Fitness trackers can function, not only as a technology of the self, but also as a technology of health. Like the "oddities of sex..." that relied "on a technology of health and pathology," the Basis fitness tracker targets that which is "[i]mbedded in bodies, becoming deeply characteristic of individuals..."[29]

The Basis website offers a reason for why self-monitoring is so popular. They write "...[t]here's good research proving that simply keeping track of what we do can significantly improve our health. Self-tracking can actually make us follow a healthier diet, sleep better and exercise more—simply by letting us know the areas we need to improve and if we are actually improving them. Fitness trackers provide this feedback in real time."[30] It is unclear from this statement what motivates this will to knowledge about the health and pathologies of our bodies. Basis assumes that it is a good thing to provide a technology that can "make" people change their behavior. The implication once again is that the drive for self-improvement is natural, and that fitness trackers are simply accommodating human nature. According to Basis and Fitbit, fitness trackers are innocuous tools that help people achieve their natural drives to better themselves. However, I would suggest that the popularity of this technology of power has more to do with the complex interplay between neoliberal governmentality, medical discourses, and biopower. Fitness trackers play a key part in the current political economy of normalized and normalizing bodies.

Basis also appeals to current studies on the benefits of fitness trackers. Their website goes on to state that "[s]tudies show that people who keep logs are the most successful at reaching their health, fitness and weight loss goals. And the more frequent and detailed the entries, the greater the success..."[31] Fitness trackers help people overcome the possibility for human error by keeping track of information better than individuals can. While studies suggest that people lose steam after several months of using fitness trackers, Basis shows that individuals can overcome this lack of motivation through competition. Competition aims to energize an internal rationality that strengthens through using fitness trackers. As an "enterprise unit" (an entrepreneur of the self), individuals need to be concerned with quality control. Thus, individuals monitoring their bodies in order to weed out their weaknesses can provide a better self-product. For Foucault, the neoliberal subject is "an entrepreneur, an entrepreneur of himself. Being for himself his own capital, being for himself his own producer, being for himself the source of [his] earnings."[32] Similar to owning a house, people must invest in up-keeping their bodies in order to get the best value when they sell themselves. According to Basis, studies "concluded that people can maintain the initial level of enthusiasm that monitoring brings by setting and regularly updating health and fitness 'goals' to foster a healthy sense of competition. And today's fitness trackers do just that."[33] Basis draws its supporting evidence from reputable medical journals, such as *The New England Journal of Medicine*, *The American Journal of Preventive Medicine*, and *The Journal of Medicine and Science in Sports and Exercise*, to establish their credibility. Basis is capitalizing on the current demand for self-monitoring technologies, while neoliberal governmentality is driving the demand for these devices, and medical discourses produce knowledge that naturalize and normalize the need for self-monitoring. Neoliberalism must work to foster a reality where humans are driven by competition rather than exchange.

Neoliberal Subjectivities in Online Fitbit Communities

Studies that have looked at the effectiveness of using fitness trackers show that individuals will continue to use these devices if they can engage in "healthy" competition and gain support from a community. Fitbit provides a blog, discussion board, and online Fitbit challenge communities to its users on its website.[34] I analyze three popular discussion forums and two Fitbit challenge groups to illustrate how individual users of the Fitbit practice neoliberal subjectivity. To maintain the integrity of the original posts, I have left them unchanged. These online communities illustrate a strategy of neoliberal governmentality aimed at decreasing their productive governing role: self-care. Lemke contends that making "individual subjects 'responsible' (and also collectives, such as families, associations, etc.) entails shifting the responsibility for social risks such as illness, unemployment, poverty, etc. and for life in society into the domain for which the individual is responsible and transforming it into a problem of 'self-care.'"[35] Thus, these ongoing discussions and groups in Fitbit's online community also provide a glimpse into the micro-political levels of neoliberal biopower.

Discussion Forum 1

The first discussion board I analyze is titled "The Financial Benefits of Using a Fitbit." This discussion board was started in 2014 by a new user of Fitbit. The user indicates that using his Fitbit Flex has saved him money, or at least will save him money, through the food log and calorie counter. I will refer to the first user as FB User 1, and subsequent users as FB User 2, FB User 3, and so on. FB User 1 provides a list of each way he thinks the Fitbit has saved him money. He begins with "1. He is eating only 1 square of an 8-ounce chocolate bar a day; he is eating less calories and fat and is saving 9 dollars a week. 2. He cut his wine intake from 1 to 2 per night to 1 glass every other night and he uses the Fitbit to help motivate him," and he goes on to write that "3. He eats less bread, including 'expensive' baguettes. 4. He consumes less chips and pretzels. 5. He saves money on gas by walking to the store instead of driving."[36] Then FB User 1 asks others to discuss how Fitbit has saved them money too. His post implies that saving money is good, and it is good because it is a form of self-discipline. The Fitbit has helped FB User 1 better control and then eliminate unwanted habits. Not only is he saving money, but he is also reducing his food intake.

FB User 1 has taken it upon himself to decrease his spending and consumption, and he assumes that both of these behavioral changes are good for him and for other people. At the end of his post, he writes that "[a]ccepting [s]olutions is your way of passing your solution onto others and improving everybody's Fitbit."[37] Here FB User 1 illustrates the neoliberal subject "who pursues his own interest, and whose interest is such that it converges spontaneously with the interest of others."[38] FB User 1 believes that what is in his best interest is also in the best interest of others since it is only natural that everyone would want to better themselves through the practices of a particular kind of self-cultivation. His practices are understood as self-cultivation or self-care. But I would suggest that they are instead processes of capital since, as I indicated earlier in this chapter, everything has been subsumed under capital. What FB User 1 is doing is not really self-cultivation or self-care, since he is actually instrumentalizing the notion of the self in order to increase his competitive edge (fitter, happier, responsible).

FB User 2 points to another form of savings through health insurance. She writes that "[m]y health insurance company is giving me 10 loyalty points per day if I reach 10,000 steps. That works out to a $20 gift card per year. Not a lot but better than nothing and when added to the other loyalty points earned for other stuff adds up."[39] Health insurance capitalizes on her human data by offering her incentives to take control of her health through the Fitbit. The incentive intensifies her desire to produce knowledge about her body. FB User 2 goes on to write: "[w]hen I walk the last 2km to work instead of taking the train to outside the door I save on one zone train fare, that's 60c/trip...I'm drinking black coffee provided by work instead of skinny lattes saving $3.80 per day."[40] Similar to FB User 1's experience, her spending behavior has changed in tandem with her increased exercise. Well-rounded "self-care" involves not only taking care of the body through better eating and exercise, but also through wiser spending choices.

Other discussion threads address how much Fitbit users are saving by no longer eating fast food. For example, FB User 3 writes: "I now eat at home almost every night after seeing how much dinner time fast foods were making me go over my calories each night. I've saved at least $20-$40 a week and that's being conservative."[41] Her comment shows that the Fitbit appeals to an underlying desire to take responsibility for the problems individuals see in their behavior, and then to make changes. Medical discourses set the norms that drive these individuals to turn a critical eye on their bodies, while neoliberalism has produced the template for an economic reality and a subject that governs itself. FB User 3 was initially driven to eat less fast food because doing so would help her reach her maximized body, but she recognizes over time how this change reduces her frivolous spending. As a good neoliberal subject, FB User 3 is invested in perfecting her human capital. She also indicates that she is teaching her son how to discipline himself; she writes that her favorite part of using the Fitbit is that "I'm teaching my son healthier habits by waiting to eat until we get home instead of grabbing fast food in a hurry. At two he's learning patience, discipline and he is trying new and healthy foods that most of the time he likes," and later she writes that "I'm not lazy anymore...And I cook now (I have to be honest and say I was just lazy about that before but the calorie count on processed foods made that decision for me.) I am really so thankful for this fitbit!"[42] She sees accountability and self-discipline as signs of good character that she must instill in her son. For Foucault, "[t]he household is a unit of production in the same way as the classical firm."[43] FB User 3 feels the need to decrease her costs so that she can increase her profit and production; if her son is disciplined and efficient her human capital can carry on through him, and he can, in time, increase his human capital too. She is in the process of transforming her unwanted body (obese and lazy) into a wanted body (thin and productive). Her thin and productive body will be more competitive, both in physical challenges and in market challenges.

FB User 3 is one among many who thank Fitbit. Neoliberalism has produced a reality where people do not resent the interference of power in their lives. Rather, they are grateful for and excited about how these technologies of power help them be better. They enjoy producing knowledge about their bodies, critically examining their bodies, and then maximizing them. She writes that "I log anything I eat BEFORE I eat it so I can see if it's a wise choice or if I'm going to slowly return that item to the shelf, lol."⁴⁴ Rather than voicing concern about the level of control this self-monitoring technology has over her, she finds it humorous that she will not eat something unless she has logged it into Fitbit's calorie tracker. Other people on the discussion board congratulate her on her self-disciplining accomplishments.

Discussion Forum 2

Another popular discussion forum is titled "You know you're a Fitbit addict when..." The first post started in 2013. I will refer to the users as YF User 1, YF User 2, and YF User 3. YF User 1 writes that you are a Fitbit addict if "You pace the house to get to a particular figure before going to bed. You sit down when recharging as you don't want to 'waste' any steps. You park further away from the shops to gain extra steps."⁴⁵ There is a certain level of pride in her obsessive use of her Fitbit. A neoliberal subject is easily governable as long as means for self-governing exist. Her post reflects the pleasure Fitbit users experience in the individual choices they make to use the Fitbit. The neoliberal subject is a subject of interest, and in order to pursue their interests they must make choices. In fact, these individuals must pursue and maximize their own interests to the edge of burn out, and in many cases, past burn out. According to Foucault, "you will see that the more they pursue their own interests the better things will be and you will have a general advantage which will be formed on the basis of the maximization of the interest of each. Not only may each pursue their own interest, they must pursue their own interest, and they must pursue it through and through by pushing it to the utmost..."⁴⁶ The playful nature of this discussion thread masks YF User 1's obsessive and intense pursuit of interest; neoliberal governmentality regulates her pursuit of interest through normalizing/maximizing discourses. It is in YF User 1's interest to be thin, fit, healthy, and highly disciplined, one could even say obsessively disciplined, in order to be a good entrepreneur of the self and to remain intelligible to neoliberal governmentalities.

She goes on to write in another post that "Yep, you're all definitely addicted! But hey, it's a good thing.... right? Mine stopped charging and was going flat within half an hour. I was getting worried. Got it replaced today and in the process lost 3 days of data but that's better than losing all my data—now that would be heartbreaking."⁴⁷ The implication here is that YF User 1's self is, in part, constructed by the knowledge she produces through self-monitoring technologies in the Fitbit. This knowledge production intensifies her sense of self, and it is clear that losing all of her data would be like losing part of herself.

YF User 2 encourages YF User 1 to continue her addictive behavior. She writes: "...you are so right. If there is a 12-step program for addicts, I don't want to even know about it. This is one addiction that I pray will stay in my life forever."⁴⁸ YF User 2 identifies a form of addiction that will help her in the pursuit of her interests. She suggests that addiction could be justified or normalized if it helps someone pursue their interests. Another user voices disappointment in herself for failing to reach an addictive level of self-monitoring; she writes "Addiction! Real close! Trying to get 3,000,000 by December 31, 2013. @ 13,000 steps a day...didn't make yesterday 13K feel bad this morning."⁴⁹ YF User 2 comforts this user by telling her not to give up, and not to feel bad, since she can reach an addictive level by just working harder. These Fitbit users regulate how others discipline and regulate their bodies too by not only sharing how they discipline and regulate their bodies, but also by offering words of encouragement for others to continue or increase their disciplining. People seek others to help them feel accountable for their actions.

A popular theme within a number of these discussion forums is competition. This theme is especially popular in the "You know you're a Fitbit addict when..." People compete with each other to prove they are more addicted to the Fitbit than others, while at the same time they provide support for others to become more addicted. YF User 3 writes: "I never take mine off except to bathe......I walk on the treadmill before it begins....I walked up the escalator....on a non-gym day, on a Saturday morning, I walk the mall three times to get over 6000 steps before breakfast......I'm competing with Lisa, Les, Dominique, and Maria in steps....I'm currently in 5th place."⁵⁰ She has produced a lifestyle that involves thinking about the moving body throughout the day. Her thread indicates that she is continually moving in order to be better than her competitors. YF User 3 modifies her own behavior according to the micro forms of domination she has over her competitors and that they have over her. She is driven by a competitive egotism that, according to medical discourses, benefit her body and the bodies of her competing Fitbit users.

Discussion Forum 3

The third discussion forum is titled "I am new to Fitbit and looking for some friends. Who would like to be partners in motivation?" FF User 1 started this forum in 2013 to find people to help her lose an additional forty pounds. She posts: "I am new to fitbit. I have lost 120 pounds in the past 2 years on my own and am now looking to lose the last 40 with support: -) Hoping to meet some great people with similar goals, because this last 40 is giving me a hard time and I do not want to become stagnant."⁵¹ Some tech-journalists have argued that fitness trackers lead individuals out of their egoistic journey to a fitter self by bringing them closer to other users in communities. Gary Wolf suggests that "...self-tracking culture is not particularly individualistic. In fact, there is a strong tendency among self-trackers to share data and collaborate on new ways of using it...people following their baby's sleep pattern with Trixie Tracker can graph it against those of other children; women watching their menstrual cycle at MyMonthlyCycles can use online tools to match their chart with others."[52] However, I would argue that these fitness tracking communities function as a way for individuals to maximize themselves. Thus, they remain a tool for the individual to increase their human capital. FF User 1 turns to the Fitbit community for friends as a motivational tool. I question whether she is looking for a true communal connection and for meaningful friendships. Neoliberalism encourages individuals to invest in themselves, expect a profit from it, and be willing to accept the risks.[53] The Fitbit community can help FF User 1 invest in herself by offering a source of motivation and support. Seeking out a community for motivation and support is still an individualistic pursuit. In another post, FF User 1 writes: "I am always here for support. I found being accountable for meeting my goals to other Fitbit community people helps me. I just need more of that."[54] Her post is consistent with a theme in this particular thread: accountability. Neoliberalism capitalizes on the kind of moral person who recognizes when they are responsible for their actions and being thin is an outward sign that a person is moral because it shows that he or she has worked hard to normalize the body. Advocating accountability is a way to lead people to their ideal state of being: thin and moral. Fitbit users seek others to regulate each other, and then discipline their bodies in order to meet these regulatory expectations.

Fitbit Challenge Groups

Fitbit also provides its users access to online fitness challenge groups. Fitbit challenge groups bring individual users together in order to compete against group members in a number of intense physiological challenges. Once members join the group, their profile becomes available to their group members. This profile availability means, in some cases, that individuals can view their group members' knowledge production, which includes weight, steps taken, water consumed, types of daily exercise, food intake, when, how well, and how long a person slept, etc. In other words, the Fitbit application provides users 24/7 access to, not only their own "self-cultivating" experience, but also to their group member's 24/7 "self-cultivation." A person can make their information private; however, group members do consistently provide other group members access to their information.

The first Fitbit challenge group is titled "Step-up or Kicked Out!" I will refer to this group as Step-up Group. This group describes itself as follows: "Step Up or Kicked Out is a 'boot' group. Get 84,000 steps/week (12K/day or better) by Friday at midnight or get kicked out for four weeks. No one wants to be kicked out, so this should motivate you to get those steps!! SuKo is also a challenge group. We run individual and team challenges to keep stepping fun—not a chore. So Step Up and join

SuKo!! Or get Kicked Out."[55] Step-up Group's association with military language works to motivate group members through intimidation and exclusion. Its expectations are higher than the Fitbit daily challenge of ten thousand steps per day. Their website provides in a prominent place a daily running graph that includes all of the group members' performances. Only the names of the people in the top three places are listed; so if group members want to be recognized they have to be in the top three places. This group provides members a place where they can practice their neoliberal subjectivity: a subjectivity that, as I suggested above, lives intensely, always on the edge of burn out. Each member can hold each other accountable for bettering their bodies, while offering regulatory support, and showing their moral character by working hard to maximize their bodies.

Step-up Group members have made their information, including their body goal, calories burned, time exercised, weight and food intake, available to their fellow fitness challengers. They are driven to share the knowledge they gain through self-monitoring by a network of biopower. Foucault argues that "power relations materially penetrate the body in depth, without depending even on the mediation of the subject's own representations. If power takes hold on the body, this isn't through its having first to be interiorized in people's consciousness. There is a network or circuit of bio-power, or somato-power..."[56] The technologies of power that develop out of the intersection between medical discourses and neoliberalism have "penetrated the body" as Foucault suggests. Fitbit users feel compelled to take charge of their body because their way of being is open to normalizing and maximizing practices. Maximization is normal. Disciplining the body then appears to be part of human nature and a sign of good character, rather than a matter of subjectivization, subjugation, or even domination. Everyday practices of the individual produce an ontology that is easily governed within neoliberalism. When a person becomes a self-monitoring cyborg by joining a fitness tracker to their body, they are doing what is natural for a neoliberal subject: increasing their human capital by living intensely.

The title of the second Fitbit challenge group is "Zombie Apocalypse Survivors...," and it describes itself as follows: "MUST walk, run or crawl ....10,000 steps every day or be eaten alive by flesh craving undead!!! We will mourn, very briefly, the underachievers! Now MOVE!!"[57] Again, group members regulate how they all discipline their bodies by producing knowledge about their bodies and making it available. The suggestion that group members will be eaten by a zombie or become a zombie if they fail to walk or run enough appeals to their fear of failing to live on the edge of burn out. It is a fear of becoming part of a general economy of bodies without the ability to maximize one's human capital: a total loss of the neoliberal self if one should fail to walk enough.

These groups show the productive relationship between power and knowledge. In "The Body of the Condemned," Foucault points to this productive relationship; he argues,

> [w]e should admit, rather, that power produces knowledge (and not simply by encouraging it because it serves power or by applying it because it is useful); that power and knowledge directly imply one another; that there is no power relation without the correlative constitution of a field of knowledge, nor any knowledge that does not presuppose and constitute at the same time power relations...the subject who knows, the objects to be known, and the modalities of knowledge must be regarded as so many effects of these fundamental implications of power-knowledge.[58]

Self-monitoring cyborgs produce a vast amount of knowledge on a daily basis, and this knowledge and power work together to continually make the neoliberal subject real. Fitbit users are both subjects of knowledge since they produce information about their bodies and objects of knowledge because it is their body that they are observing.

Conclusion: The Desperation Haunting "Self-Cultivation"

Haraway calls for a refusal of an "anti-science/technology metaphysics" in order to reconstruct a reality that recognizes the social relations of science and technology and the connections and myriad forms of communication between persons.[59] For Haraway, the cyborg would be the subject of this reality, and it would challenge the dualisms of dominant discourses. However, neoliberalism is not hostile to science, nor has it demonized technology. On the contrary, the self-monitoring cyborg represents a tight coupling between the normalizing sciences of medical discourses and neoliberalism. Neoliberal subjects embrace technology's ability to enhance their competitive edge. People develop intimate relationships with their self-monitoring devices, suggesting that these devices become part of their being as they intensify their ability to practice a particular kind of subjectivity. As enhanced neoliberal subjects, self-monitoring cyborgs perpetuate dualistic thinking by constantly focusing on normal/abnormal, moral/immoral, lazy/productive, and competitive/apathetic binaries.

In *No Speed Limit*, Shaviro argues that self-cultivation is unthinkable, perhaps unintelligible, in neoliberalism because it requires a "reflexive turning inward," which is for Shaviro "the opposite of self-branding, where I stylize myself in order to market myself, to be an entrepreneur of myself, and to increase the value of my 'human capital.'"[60] I am not so sure that unintelligibility is the right way to think about how notions of self-cultivation might function or persist within neoliberalism. Many of the tech companies that produce fitness trackers and fitness applications utilize notions of self-cultivation in tandem with ideas about living intensely. It is as though living intensely is a form of self-cultivation for these companies and for self-monitoring cyborgs. This is, of course, not to suggest that living intensely as self-cultivation is actually self-cultivation since, as I argued earlier in this chapter, self-cultivation is not dependent on its instrumental value. It is about caring for the self for its own sake. Rather, neoliberal governmentalities, in tandem with nodes of power/knowledge, benefit from the persistence of notions of self-cultivation that hide the true nature of practices of self-care: namely that all aspects of life are now subsumed under capital, or that all aspects of life are now functions of capital/ism. People still think they are engaging in practices of self-cultivation when, in reality, they are engaging in the dissolution of the self, since they are in fact machinic functions of capitalism. This quasi-life of struggling to live intensely, to emulate the privileged, is not just a by-product of neoliberalism, as Crary may suggest. It is not something that comes as an unintended effect of neoliberalism, but rather is very much what the neoliberal subject is and does. This kind of subject is endlessly governable under neoliberalism and thrives under these conditions since it valorizes and desires the assumption of risk and the responsibility of its choices for successfully competing. When the neoliberal subject reaches burn out, it is only "natural" that it blames itself. Neoliberal quasi-life, a life that struggles to endure, and a life that is not quite living and not quite dead, is a notion that I take up in Chapter 3. In this next chapter, I consider how we might rethink biopower apart from its focus on "outright extinction,"[61] and I examine the ways in which Foucault's analytic more so than Achille Mbembe's concept of necro-politics may help us to continue to make life strange under neoliberalism.

Notes

1. Ibid., 9.
2. Thomas Lemke, "The Risks of Security: Liberalism, Biopolitics, and Fear," in *The Government of Life: Foucault, Biopolitics, and Neoliberalism*, eds. Vanessa Lemm and Miguel Vatter (New York: Fordham University Press, 2014), 68.
3. Jonathan Crary, *24/7: Late Capitalism and the Ends of Sleep* (London: Verso, 2013), 9.
4. Steven Shaviro, *No Speed Limit: Three Essays on Accelerationism* (Minneapolis: University of Minnesota Press, 2015): 28–29.
5. William Gibson, *Neuromancer* (New York: Ace, 1984), 7.
6. Jeffrey T. Nealon, *Post-Modernism or, The Cultural Logic of Just-in-Time Capitalism* (Stanford: Stanford University Press, 2012).
7. Robin James. "Loving the Alien," *The New Inquiry*, 2012, https://thenewinquiry.com/loving-the-alien/.
8. Huawei is a technology company that produces fitness trackers.

9. "Mission Statement," *Huawei*, accessed on June 26, 2018, https://consumer.huawei.com/en/about-us/.
10. Shaviro, *No Speed Limits*, 26.
11. Donna Haraway, "A Cyborg Manifesto: Science, Technology, and Socialist-Feminism in the Late-Twentieth Century," *Simians, Cyborgs and Women: The Reinvention of Nature* (New York: Routledge, 1990), 163.
12. Gary Wolf, "Know Thyself: Tracking Every Facet of Life, from Sleep to Mood to Pain, 24/7/365," *Wired*, June 22, 2009. http://archive.wired.com/medtech/health/magazine/17-07/lbnp_knowthyself?currentPage=all.
13. Michel Foucault, *The Birth of Biopolitics: Lectures at the Collège de France, 1978–1979* (New York: Palgrave MacMillan, 2004), 219–223.
14. Ryan Holmes, "Do You Fitbit? Why Health Trackers Are So Hot Right Now," *Linked In*, July 9, 2014, https://www.linkedin.com/pulse/20140709164338-2967511-do-you-fitbit-why-health-trackers-are-so-hot-right-now.
15. Ibid., 223.
16. Michel Foucault, "*The Birth of Biopolitics*," 228.
17. Albert Sun, "The Monitored Man," *The New York Times*, March 10, 2014, http://well.blogs.nytimes.com/2014/03/10/the-monitored-man/.
18. Michel Foucault, "The Right of Death and Power Over Life," in *The Foucault Reader*, trans. Paul Rabinow (New York: Pantheon, 1984), 262.
19. Ibid., 262.
20. Carol Ewing Garber et al., "Quantity and Quality of Exercise for Developing and Maintaining Cardiorespiratory, Musculoskeletal, and Neuromotor Fitness in Apparently Healthy Adults: Guidance for Prescribing Exercise," *The Journal of Medicine and Science in Sports and Exercise* 43, no. 7 (2011).
21. Michel Foucault, "The Politics of Health in the Eighteenth Century," in *The Foucault Reader*, trans. Paul Rabinow (New York: Pantheon, 1984), 283.
22. Ibid., 277.
23. Ibid., 277.
24. Haraway, "A Cyborg Manifesto," 150.
25. "Who We Are," *Fitbit*, accessed June 25, 2018, https://www.fitbit.com/about.
26. "Mission Statement," *Fitbit*, accessed June 25, 2018, https://www.fitbit.com/about.
27. Michel Foucault, *The History of Sexuality: Volume 1* (New York: Vintages Books, 1990), 86.
28. "Mission Statement," *Basis*, accessed May 4, 2015, https://www.mybasis.com/basis-fitness-tracker/.
29. Foucault, *The History of Sexuality*, 44.
30. "Mission Statement," accessed May 4, 2015, https://www.mybasis.com/basis-fitness-tracker/. Basis no longer produces BasisPeak, its primary fitness tracker. While it now focuses on sleep tracker, its BasisPeak Mission statement is indicative of many of the corporate mission statements for fitness trackers.
31. Ibid.
32. Foucault, "*The Birth of Biopolitics*," 226.
33. "Mission Statement," https://www.mybasis.com/basis-fitness-tracker/.
34. Fitbit no longer offers web-based community discussions. Community activity now occurs within the Fitbit app. However, these web-based community discussions are indicative of the kinds of discussions that continue within Fitbit communities.

35. Thomas Lemke, "Foucault, Governmentality, Critique," presentation, The Rethinking Marxism Conference (University of Amherst (MA), September 21–24, 2000).
36. "The Financial Benefits of Using the Fitbit," accessed May 1, 2015, https://community.fitbit.com/t5/Share-Your-Story/financial-benefits-of-using-a-FitBit/td-p/389416/highlight/true. Again, Fitbit no longer offers web-based community discussions. Community activity now occurs within the Fitbit app. However, these web-based community discussions are indicative of the kinds of discussions that continue within Fitbit communities.
37. Ibid.
38. Foucault, *"The Birth of Biopolitics,"* 270.
39. "The Financial Benefits of Using the Fitbit," accessed May 1, 2015, https://community.fitbit.com/t5/Share-Your-Story/financial-benefits-of-using-a-FitBit/td-p/389416/highlight/true.
40. "The Financial Benefits of Using the Fitbit," https://community.fitbit.com/t5/Share-Your-Story/financial-benefits-of-using-a-FitBit/td-p/389416/highlight/true.
41. Ibid.
42. "The Financial Benefits of Using the Fitbit," https://community.fitbit.com/t5/Share-Your-Story/financial-benefits-of-using-a-FitBit/td-p/389416/highlight/true.
43. Foucault, *"The Birth of Biopolitics,"* 244.
44. "The Financial Benefits of Using the Fitbit," https://community.fitbit.com/t5/Share-Your-Story/financial-benefits-of-using-a-FitBit/td-p/389416/highlight/true.
45. "You Know You're a Fitbit Addict When," accessed May 1, 2015, https://community.fitbit.com/t5/Share-Your-Story/You-know-you-re-a-Fitbit-addict-when/m-p/3773/highlight/true#M116.
46. Foucault, *"The Birth of Biopolitics,"* 275.
47. "You Know You're," https://community.fitbit.com/t5/Share-Your-Story/financial-benefits-of-using-a-FitBit/td-p/389416/highlight/true.
48. Ibid.
49. Ibid.
50. Ibid.
51. "I am new to Fitbit and looking for some friends. Who would like to be partners in motivation?" accessed May 1, 2015, https://community.fitbit.com/t5/Share-Your-Story/I-am-new-to-fitbit-and-am-looking-for-some-friends-Who-would/td-p/14729/highlight/true.
52. Gary Wolf, "Know Thyself: Tracking Every Facet of Life, from Sleep to Mood to Pain, 24/7/365," *Wired Magazine* June 22, 2009.
53. Foucault, *"The Birth of Biopolitics,"* 252.
54. "I am New to Fitbit," https://community.fitbit.com/t5/Share-Your-Story/I-am-new-to-fitbit-and-am-looking-for-some-friends-Who-would/td-p/14729/highlight/true.
55. "Step Up or Kicked Out," accessed May 1, 2015, https://www.fitbit.com/group/22R4DN.
56. Michel Foucault, "The History of Sexuality," *Power/Knowledge* (New York: Vintage Books, 1980): 186.
57. "Zombie Apocalypse Survivors," accessed May 1, 2015, https://www.fitbit.com/group/224KLW.
58. Michel Foucault, "The Body of the Condemned," in *Discipline and Punish: The Birth of the Prison*, trans. Alan Sheridan (New York: Pantheon, 1977), 27–28.

59. Haraway, "A Cyborg Manifesto," 181.
60. Shaviro, *No Speed Limits*, 51.
61. Kathryn Yusoff. "An Interview with Elizabeth Povinelli: Geontopower, Biopolitics and the Anthropocene," By Mathew Coleman and Kathryn Yusoff. *Theory, Culture & Society* 34, no. 2–3 (2017), 169–185.

Bibliography

Crary, Jonathan. *24/7: Late Capitalism and the Ends of Sleep*. London: Verso, 2013.

"Financial Benefits of Using the Fitbit," Accessed May 1, 2015, https://community.fitbit.com/t5/Share-Your-Story/financial-benefits-of-using-a-FitBit/td-p/389416/highlight/true.

Foucault, Michel. "The Body of the Condemned," in *Discipline and Punish: The Birth of the Prison*, trans. Alan Sheridan. New York: Pantheon, 1977.

—., "The History of Sexuality," in *Power/Knowledge*. New York: Vintage Books, 1980.

—., "The Politics of Health in the Eighteenth Century," in *The Foucault Reader*. Translated by Paul Rabinow. New York: Pantheon, 1984.

—., "The Right of Death and Power Over Life," in *The Foucault Reader*. Translated by Paul Rabinow. New York: Pantheon, 1984.

—., *The History of Sexuality*. New York: Vintage Books, 1990.

—., *The Birth of Biopolitics: Lectures at the Collège de France, 1978–1979*. New York: Palgrave MacMillan, 2004.

Garber, Carol Ewing et al. "Quantity and Quality of Exercise for Developing and Maintaining Cardiorespiratory, Musculoskeletal, and Neuromotor Fitness in Apparently Healthy Adults: Guidance for Prescribing Exercise," *The Journal of Medicine and Science in Sports and Exercise* 43, no. 7 (2011).

Gibson, William. *Neuromancer*, New York: Ace, 1984.

Haraway, Donna. "A Cyborg Manifesto: Science, Technology, and Socialist-Feminism in the Late-Twentieth Century," *Simians, Cyborgs and Women: The Reinvention of Nature*. New York: Routledge, 1990.

Holmes, Ryan. "Do You Fitbit? Why Health Trackers Are So Hot Right Now," *Linked In*, July 9, 2014, https://www.linkedin.com/pulse/20140709164338-2967511-do-you-fitbit-why-health-trackers-are-so-hot-right-now.

"I am new to Fitbit and looking for some friends. Who would like to be partners in motivation?" Accessed May 1, 2015, https://community.fitbit.com/t5/Share-Your-Story/I-am-new-to-fitbit-and-am-looking-for-some-friends-Who-would/td-p/14729/highlight/true.

James, Robin. "Loving the Alien." *The New Inquiry*. 2012. https://thenewinquiry.com/loving-the-alien/.

Lemke, Thomas. "Foucault, Governmentality, Critique," (presentation, The Rethinking Marxism Conference, University of Amherst (MA), September 21–24, 2000).

"Mission Statement," *Basis*, Accessed May 4, 2015, https://www.mybasis.com/basis-fitness-tracker/.

"Mission Statement," *Fitbit*, Accessed May 1, 2015, https://www.fitbit.com/about.

"Mission Statement," *Huawei*, Accessed June 26, 2018, https://consumer.huawei.com/en/about-us/.

Nealon, Jeffrey T., *Post-Modernism or, The Cultural Logic of Just-in-Time Capitalism*, Stanford: Stanford University Press, 2012.

Shaviro, Steven. *No Speed Limit: Three Essays on Accelerationism*, Minneapolis: University of Minnesota Press, 2015.

"Step Up or Kick Out," Accessed May 1, 2015, https://www.fitbit.com/group/22R4DN.

Sun, Albert. "The Monitored Man," *The New York Times*. March 10, 2014. http://well.blogs.nytimes.com/2014/03/10/the-monitored-man/.

"You Know You're Addicted to Fitbit When," Accessed May 1, 2015. https://community.fitbit.com/t5/Share-Your-Story/financial-benefits-of-using-a-FitBit/td-p/389416/highlight/true.

"Who We Are," Fitbit, Accessed May 1, 2015, https://www.fitbit.com/about.

Wolf, Gary. "Know Thyself: Tracking Every Facet of Life, from Sleep to Mood to Pain, 24/7/365," *Wired*. June 22, 2009. http://archive.wired.com/medtech/health/magazine/17-07/lbnp_knowthyself?currentPage=all.

"Zombie Apocalypse Survivors," Accessed May 1, 2015, https://www.fitbit.com/group/224KLW.

3 Subtle State Killing as a Mode of Neoliberal Governmentality

Introduction

This chapter seeks to disentangle Foucault's work on biopolitics from Achille Mbembe's "Necropolitics." It argues that the "death making" of neoliberal governmentalities can still be understood in terms of Foucault's analytic. Foucault's biopolitics is often conflated with Mbembe's necropolitics (they both address genocide and concentration camps), even though Mbembe criticized Foucault's biopower for its inability to account for the ways spectacular modes of killing function. Thus, the goal of this chapter is, in part, to show that Foucault's biopolitics can do more than theorize a genealogy of biological racism and genocide. That is, under neoliberalism, biopolitics does not eliminate populations solely through the production of biological caesuras predicated on race, and this elimination process does not primarily operate through spectacular and large-scale death-making, as the murder function of the state. Rather, as I suggested in Chapter 1 and in the Introduction, neoliberal biopolitics functions, in part, through the marketization of space and generalized conditions of competition. And, as I suggested in Chapter 2, the self-monitoring cyborg is one example of the kinds of neoliberal subjects that compete to live, and if they fail to compete must endure burnout: a quasi-life, a life that is not quite living and not quite dead. Necropolitics' focus on spectacular killing may miss the kind of quasi-life that emerges out of neoliberal biopolitics and its marketized spaces. Necropolitics may miss the ways in which neoliberal subjects continually live on the edge of burn out and the ways in which we are all incited to live the most intense lives. In these marketized spaces, when populations of individuals fail to compete (they do not accurately assess risk, or they unsuccessfully maximize their human capital, or they are incapable of enduring or preventing chronic illnesses), it seems "natural" that they should deal with the consequences of their failure to compete. These consequences, whether they be slow-burning death from chronic illnesses, homelessness, physiological and emotional stress, etc., are understood as the responsibility and the fault of the individual, not of the state or of a particular political/economic/social system. Thus, if part of the goal of this project is to highlight the ordinary/taken-for-granted deaths and death-making that permeate the marketized spaces of competition under neoliberalism, disentangling Foucault's biopolitics from necropolitics's focus on "engineered elimination"[1] is a necessary step. Biopolitics without its zero-sum focus may turn our attention to the subtle and deadly ways in which neoliberalism functions. My aim is not to defend Foucault or suggest that we must focus on ordinary deaths at the expense of spectacular modes of killing. Rather, the point is that death-making under neoliberalism exists within a spectrum (it is not spectacular death or ordinary death), and focusing on spectacular deaths, as necropolitics tends to do, often misses the other subtler modes of death-making that are very much a part of neoliberalism.

Mbembe poses several questions about the sufficiency of Foucault's "biopower" as an analytic for determining and interpreting contemporary examples of when the political uses murder to achieve its ends. Essentially, Mbembe is asking whether biopower can help us to account for the ways state forces and sub-state forces kill (whether it is during times of war, moments of resistance, or fighting against terrorism) the people they have deemed to be enemies. Mbembe's answer is that biopower is limited in what it can tell us about the ways modern sovereignty, the political, states, and sub-states subjugate life to the power of death. This subjugation of life to the power of death is what Mbembe calls necropolitics. Necropower, for Mbembe, describes the spatial and temporal ordering of contemporary colonial spaces, which involve territorial fragmentation, medieval siege warfare, and continual violence and killing.[2] Mbembe offers necropower and necropolitics as correctives to what he sees as biopower's inability to account for the irrational, excessively cruel, and spectacular forms of killing in the colonies.

Furthermore, Mbembe relates Foucault's biopower to the state of exception and the state of siege in order to show that biopower is insufficient for making sense of how the right to kill has normalized the state of exception and the relation of enmity. The suggestion here is that Foucault's biopower cannot account for the spatial and temporal logic of modern warfare, nor can it tell us about what Mbembe sees as the domination of a politics of death, and of spectacular death at that, in modern statehood. Mbembe implies that Foucault's notion of biopower cannot make sense of the connections between politics and death in systems that operate within a permanent state of emergency.[3] According to Mbembe, death and power function under a different logic within the permanent state of emergency compared to the way they do within biopower.

Mbembe's essay "Necropolitics" was published in 2003, and it (along with Giorgio Agamben's and Roberto Esposito's writings, among others) has since been influential in shaping the way other scholars have interpreted the relationship between life and death in biopolitics. That is, biopolitics' "make live requirement" is understood as the flipside of spectacular modes of "letting die." Or, once again, biopolitics is an insufficient analytic for understanding strong state killing and states of exception. Necropolitics has been criticized for its inability to see beyond biopolitics' "zero point" (e.g., death).[4] However, if, as scholars of biopolitics argue, killing certain populations is a positive condition of biopower,[5] then biopower can still account for relations of enmity, modern warfare, and the irrational, excessively cruel, and spectacular forms of killing in colonies. More importantly, biopolitics is also no longer an insufficient analytic for understanding ordinary suffering as a mode of state killing. Unlike Mbembe, I suggest that biopower does not need a state of exception to justify killing, if by definition, biopower kills to make live certain populations. Thus, I wish to argue that we do not need a flipside or an addendum to biopower, such as necropolitics or thanatopolitics,[6] since biopower can itself account for subtle and overt modes of state killing.

Furthermore, while Mbembe has been useful for conceptualizing the spatio-temporality of "large-scale killings"[7] and permanent extrajuridical killing, his necropolitics misses the paradoxical nature of biopower and overlooks the subtle modes of state killing that capitalize on ordinary suffering under neoliberalism. In this chapter, I address a number of problems with Mbembe's understanding of sovereignty, with his definition of biopower, and with his understanding of the function of racism within biopower. His criticism of biopower is predicated upon a conflation of sovereignty with biopower, and it results from a misreading of Foucault's understanding of sovereignty. Mbembe's necropower, although useful in thinking about the creation of death-worlds in the context of contemporary forms of warfare, including colonial spaces, is thus limited in what it can offer to an analysis of the spatio-temporality of neoliberalism. Massive-scale killing and overt state killing are *not* the modus operandi of the biopolitics of neoliberal spatio-temporality. I begin this chapter with an analysis of Mbembe's definition of sovereignty and his conflation of sovereignty with biopower. Then, I address the problems with Mbembe's definition of biopower and with the way Mbembe uses Foucault's definition of racism. Finally, I consider the weaknesses of Mbembe's necropolitics/necropower in making sense of the spatio-temporality of neoliberalism.

Sovereignty, biopower, racism

Mbembe begins "Necropolitics" with the following point about sovereignty: "the ultimate expression of sovereignty resides, to a large degree, in the power and the capacity to dictate who may live and who must die..."[8] He indicates in a footnote that his approach to the question of sovereignty builds on Foucault's notion of sovereignty from his lecture "*Society Must Be Defended*." However, Mbembe's assumption that the ultimate expression of sovereignty lies in its power and capacity to determine "who may live" and "who must die" is not entirely consistent with Foucault's critique of sovereignty. Foucault uses the phrases "what must live" and "what must die" when he answers the question "what is racism?" State racism is for Foucault a way for biopower to justify the right to kill by deciding who must die and who must live.[9] But state racism and sovereignty are not the same. I will return to this point below.

For Foucault, the "[s]overeign power's effect on life is exercised only when the sovereign can kill...It is the right to take life or let live."[10] Mbembe's position that sovereignty dictates who may live suggests a more active form of power over living. Foucault's position is instead that sovereignty has an indirect power over living. Foucault writes that, "the right of life and death is a dissymmetrical one,"[11] with letting live meaning, in part, refraining from killing. But this is not a power based on "generating forces, making them grow, and making them live."[12] The ultimate expression of sovereignty, an expression of its absolute power, would be through the sovereign's right to kill, the right of the sword, and the right to put to death. According to Foucault, this "was the moment of the most obvious and most spectacular manifestation of the absolute power of the sovereign."[13] To kill, then, does not necessarily constitute the limits of sovereignty, since it is largely through killing that the sovereign exercises its power.

At times, Mbembe simply conflates sovereignty with biopower. For example, Mbembe suggests that "to exercise sovereignty is to exercise control over mortality and to define life as the deployment and manifestation of power."[14] Mbembe summarizes Foucault's biopower based on the terms he introduces at the beginning of his essay (the right to kill, to allow to live, or to expose to death), which is for Mbembe "that domain of life over which power has taken control."[15] But in the case of Foucault, the right to kill and to allow to live refers to "sovereignty's old right," "the power of the sword," or the old sovereign's right to kill.[16] The mode/system of biopower introduces a paradox in its hold over life, which is that killing or the right to kill goes against the imperative to make live and to ensure the survival of a population. Biopower's emergence as a type of power does not mean that sovereign power completely disappears or that biopower is a new form of sovereignty. Mbembe's use of mortality and "life as the deployment and manifestation of power" are elements of biopower, not of sovereignty. Yet, Mbembe writes that to do one is to do the other. Mbembe's definition of sovereignty actually reads like Foucault's definition of biopower, and specifically like the beginning of Foucault's definition of racism. When Foucault refers to mortality in "*Society Must Be Defended*," he does so when explaining biopower. Biopower concerns itself with a number of processes, including mortality rates.

Mbembe's definition of biopower (that domain of life over which power has taken control) is borrowed from Foucault's definition of racism. It is in this context that Foucault writes about "a break into the domain of life that is under power's control."[17] Biopower is implied in this sentence but not succinctly defined. Foucault offers a more succinct, yet slightly different, definition of biopower in *The History of Sexuality, Volume I*. It is different from the definition he provides in "*Society Must Be Defended*" because it includes both anatomo-politics, which was previously understood within the context of disciplinary power, and biopolitics as a series of techniques of biopower. In *The History of Sexuality, Volume I*, Foucault writes that biopower developed in two forms that are not oppositional, but instead, are entangled so that they strengthen each other even though they may at times seem to be contradictory. Anatomo-politics is "centered on the body as a machine; its disciplining, the optimization of its capabilities, the extortion of its forces, the parallel increase of its usefulness and its docility, all this was ensured by the procedures of power that characterized the disciplines."[18] It is also individualizing in that it is concerned with discipline at the level of the individual body. The other technique of biopower is biopolitics. Biopolitics is "focused on the species body, the body imbued with the mechanics of life and the serving as the basis of the biological process: propagation, births, mortality, the level of health, life expectancy and longevity."[19] Biopolitics is massifying in that it is concerned with humans at the level of the population. Foucault writes in *The History of Sexuality, Volume I* that both of these techniques of power constitute biopower. Biopower's main purpose is "no longer to kill, but to invest life through and through."[20] It is the power to "make" live and "let" die.[21] Biopower's mode is regulatory, normalizing, and still disciplinary. Killing, then, becomes a way of cleansing and ensuring the health of certain populations, but it is not the right of the sovereign to kill.

Thus, Mbembe's definition of biopower and his reading of Foucault do not do justice to the complexity of and variations in Foucault's biopower. This may pose a problem for Mbembe's assessment of whether biopower can sufficiently account for contemporary forms of warfare. Mbembe writes that biopower functions through "dividing people into those who must live and those who must die,"[22] and through the "subdivision of the population into subgroups, and the establishment of a biological caesura between the ones and the others."[23] Both of these processes of division once again refer primarily to what Foucault calls racism. According to Mbembe, racism is mostly a technology "aimed at permitting the exercise of biopower, 'that old sovereign right of death.'"[24] It is unclear what Mbembe means here. If he is suggesting that biopower is "that old sovereign right of death," then this is another conflation of sovereignty with biopower. In "*Society Must Be Defended*," Foucault argues that one way a normalizing society or the power of normalization kills is through biological racism. However, Mbembe's definition of racism as it functions within biopower does not account for the second way racism functions.

For Mbembe, the purpose of racism within biopower is to justify the right to kill and to establish a positive relation between life and death. On the other hand, for Foucault, within biopower, killing once again functions as a way to ensure that a population continues to live. That is, racism allows for the purification, health, and survival of one population by killing an inferior race, a degenerate race, or a dangerous race. The more the "good" race kills other inferior races, the healthier the good race becomes. This killing is then a biological relationship.[25] Thus, unlike what Mbembe seems to suggest, racism is not just about regulating the distribution of death and enabling the murder function of the state. This characterization reads too much like the old sovereign right to kill. Rather, racism is once again primarily about making a population healthier, purer, or more likely to live longer. Racism is instrumentalized by biopower as a mechanism designed to improve the health of a population. I have drawn attention to these ambiguities here in order to suggest that Mbembe's critique of biopower is located within a logic of sovereignty that misses the Foucaultian points about biopower. I further suggest that the logic of biopower can account for the modes of spectacular death that Mbembe addresses.

Mbembe's Assessment of Biopower

Mbembe associates biopower with a rational and normative understanding of sovereignty, which limits how we can interpret the right to kill. This form of sovereignty misses the way contemporary forms of the right to kill have normalized the state of exception and the relation of enmity. The power that functions in this right to kill utilizes "exception, emergency, and a fictionalized notion of the enemy,"[26] and Mbembe suggests that this power is not limited to state power. This extra note about state power is a reference to what Mbembe sees as a limitation of biopower in that it seems that, for him, it is exercised within/by the state.[27] It is clear to Mbembe that the sovereign right to kill and biopower are located within the functioning of the modern state, and thus are constitutive of state power primarily. For Mbembe, biopower and its ties to a sovereignty that works within the law cannot fully account for a politics of death within a state of emergency where the law is suspended. Mbembe draws a distinction between sovereignty that is dominated by reason and the norm and a modality of sovereignty that is dominated by states of exception, terror, and extermination. In other words, for Mbembe, more recent political criticisms, including Foucault's, have differentiated politics from war whereas Mbembe wishes to "imagine politics as a form of war."[28] But, once again, Foucault characterizes biopower as a mode of power that does *not* function through state power. For Foucault, biopower points to the complexities of power relations as they function in and through bodies, populations, power/knowledge regimes, and governmentalities.

In other words, biopower for Foucault is not entirely understood politically in terms of macro-relations. Biopower is not primarily about clear state actors or sub-state actors making live and letting die; it is not about top-down power.

Later, Mbembe argues that the Nazi regime is an example of one among many "early and late modern" conceptions of sovereignty. This again suggests that Foucault's biopower has been limited by its application to older forms of sovereignty, as well as by its consideration of the logic of modernity, which further is rooted in Western European history and partly attributable to Hobbes' perspective on sovereign power. In some cases, modernity/terror is expressed for Mbembe through an ancient "passion for blood" and "notions of justice and revenge."[29] In other moments, the killing of the enemy of the state is based on what Mbembe calls more "intimate, lurid, and leisurely forms of cruelty."[30] And later Mbembe argues that any consideration for the ascendance of modern terror should address slavery. Slavery is, according to Mbembe, "one of the first instances of biopolitical experimentation."[31] Mbembe is also critical of Foucault's genealogy of power because its acceptance of a normative form of sovereignty ignores the basic nature of sovereignty, which for Mbembe is a sovereignty that "consists fundamentally in the exercise of a power outside the law and where 'peace' is more likely to take on the face of a 'war without end.'"[32] Foucault does briefly touch on biopower and colonization in "*Society Must Be Defended.*" Racism offers a way to understand colonization as a confrontation understood in terms of evolutionism. According to Foucault, racism develops "with colonization…with colonizing genocide."[33] But, for Mbembe, this does not sufficiently address the spatial and temporal realities of colonial towns, of sovereignty in colonial spaces, or of the way the right to kill may function in a permanent state of exception.

Mbembe is also concerned with how the sovereign exception subjectivizes individuals in colonial spaces. Since the colonies are spaces in the state of exception, law and order are suspended. The subject of sovereign exception is the savage. Savage life and animal life are similar to Agamben's bare life here in that they are recognized as the lives of "'natural' human beings who lack the specifically human character, the specifically human reality,"[34] and therefore are excluded from the law and can be killed with impunity. But where Agamben considers the Nazi camp as the "fundamental biopolitical paradigm of the West,"[35] Mbembe extends this analysis to an examination of contemporary colonial spaces. The sovereign right to kill in these colonial spaces "is not subject to any rule in the colonies," including legal and institutional rules. However, what Mbembe considers the sovereign right to kill in Foucault's biopower is still subject to the law. As Foucault suggests, after the emergence and function of biopolitics and anatomo-politics, sovereignty was integrated with a "juridical apparatus."[36] Part of the problem with Mbembe's analysis of Foucault's biopower is that Mbembe conflates killing in biopower with the sovereign right to kill. Killing, whether it is overt or subtle state killing, in modes of biopower does not need to be established by the logic of the law or rights. Rather, if, as I suggested earlier, killing (including overt and subtle, as well as state and non-state, forms of killing) is a positive condition of biopower, then killing can function as a natural way of maintaining desired or even healthy populations. As I have been suggesting all along, neoliberalism needs and produces subjects that want insecurity, that want to live on the edge of burn out, that want to live the most intense lives. Insecurity, risk, intensified individualism, and competing to live are forms of neoliberal biopolitics. It is, in part, through insecurity, risk, intensified individualism, and competing to live, etc. that neoliberalism functions, that it makes individuals live. Populations of individuals eliminate themselves from competitive populations of individuals (the desirable populations) or they willingly and even at times celebrate enduring the living conditions that neoliberalism produces. Neoliberalism does not necessarily need the law or sovereignty to kill or to eliminate undesirable populations.

Is Necropolitics Really About Sovereign Power?

For Mbembe, the notion of necropolitics must be deployed because, according to him, necropolitics offers an analytic that can account for contemporary terror formations where irrational, excessively cruel, and spectacular forms of killing occur. Here, terror formations are the topographies of death-making that exist in a state of exception, a state of emergency, or a state of siege. These terror formations are based on complex interactions between biopower, the state of exception, and the state of siege. Mbembe's inclusion of the state of exception and the state of siege shows that he does not consider biopower sufficient in what it can explain about the sovereign right to kill in not only contemporary colonial spaces, but also in contemporary forms of resistance and terrorism. Before he offers a clearer definition of necropolitics, Mbembe examines the colony and the apartheid regime as precursory forms of necropolitical spaces. It is within the colonial world that we find "the first syntheses between massacre and bureaucracy, that incarnation of Western rationality."[37] Mbembe suggests that while Foucault does characterize war in the mode of biopower as the bloodiest yet, writing that "massacres have become vital,"[38] he does not provide an analysis of the practical and material effects of war, nor does he grasp the particular character of the spatial and temporal formations of war. Mbembe's necropolitics recognizes that "wholesale slaughter" is no longer between two "civilized" states, but rather that it occurs within terror formations, such as the colony, apartheid, or contemporary modalities of colonial occupation. For Mbembe, necropolitics can account for the spatialization of colonial occupation in ways that perhaps biopower cannot. However, here again, Mbembe fails to do justice to Foucault's analysis of colonized spaces. Foucault examines the ways in which the spaces of the colonized and the spaces of the colonizer are co-productive. In "*Society Must Be Defended*," expanding upon Jean-Paul Sartre's "Preface" to Fanon's *The Wretched of the Earth*,[39] Foucault characterizes this co-productive relationship as a "boomerang effect," which highlights the ways the techniques of colonial militarism, such as modes of hyper surveillance and often forms of colonial violence, reverberated back into western urban discourses and security practices. This point suggests that Foucault was not only interested in western imperialism independent of its colonial practices; rather, Foucault was, in part, concerned with how securitization functioned as a technique of biopower. While Foucault does not consider some of the necropolitical practices Mbembe examines, this does not mean that biopower cannot account for these modes of massive scale state and substate killing. Again, Foucault's biopower is not limited to a particular form of sovereign power.

Although Foucault does not offer a detailed account of the spatialities of biopower in "*Society Must Be Defended*," he does consider the spatialization of disciplinary power within urban spaces. The subject of disciplinary power is made into a subject, in part, through its relationship to, interactions with, and physical positioning within disciplinary spaces, such as barracks, prisons, mental hospitals, the doctor's office, and or even the home. If we consider anatomo-politics as part of biopower, then this disciplinary space does reflect a spatiality of biopower. Furthermore, Foucault makes the case in "*Society Must Be Defended*" that, as a technique of biopower, biopolitics does partly take its knowledge from "the effects of the environment."[40] Biopolitics is concerned with the problems that emerge out of humans living in certain spaces; that is, it accepts that humans affect their environment and that the environment has an effect on humans, and, as such, that mortality, birth rates, and the overall health of human populations are shaped by their environmental milieus. For Mbembe, biopolitics understands the relationship between humans and their physical environment only in terms of life and living, which allegedly cannot account for the politics of death around which colonial occupation is ordered. Here again, we see how such a reading of biopolitics is insufficient.

The town of colonial occupation is the epitome of how necropower works. Mbembe borrows from Fanon's description of the colonial township and from his understanding of colonial space. For Fanon, the colonizer sees the town as belonging to the colonized other. In the eyes of the colonizer, these towns are places with bad reputations, especially for sexual promiscuity, and they are populated by savages, degenerates, and criminals. Because of its ill repute and savage occupants, the fact of their birth, death, and quality of life do not matter.[41] The township is ordered in a way that allows the colonizer to define and dictate "who matters and who does not, and who is disposable and who is not."[42] Thus, the colonial township reflects the spatial ordering of necropower, and it shows a particular form of sovereignty. For Mbembe, this sovereignty does not align with the sovereign right to kill that intersects with biopower.

Mbembe further identifies necropower as a terror formation. According to Mbembe, contemporary forms of colonial occupation, or what Mbembe calls late-modern colonial occupation, are different from early-modern modalities of occupation because of a particular combination of the disciplinary, the biopolitical, and the necropolitical. Mbembe examines three dominant characteristics of necropower that he sees in the occupation of Gaza and the West Bank, which include territorial fragmentation, vertical sovereignty, and medieval siege warfare. Through necropower, these occupied spaces are transformed into a permanent state of siege, where individuals can be killed by anyone and civil order is destroyed. Killing does not recognize the difference between internal and external enemies. In addition to late-modern colonial occupation, Mbembe addresses necropower within contemporary warfare. He alludes to Foucault's assessment of war when he argues that contemporary wars can no longer be understood via Carl von Clausewitz's "instrumentalism." In *"Society Must Be Defended,"* Foucault locates war within the state. He also analyses Clausewitz's position that war is a continuation of politics by other means, even though Foucault further clarifies that this thesis predates Clausewitz. Foucault follows this thesis to the development of the "army as institution," and to the idea of a society "traversed by the relations of war."[43] Mbembe turns to these insights by Foucault to imply that Foucault's understanding of war and perhaps even of how biopower functions within war is limited. Mbembe writes that, "an important feature of the age of global mobility is that military operations and the exercise of the right to kill are no longer the sole monopoly of states, and the 'regular army' is no longer the unique modality of carrying out these functions…" and then he indicates that "[i]nstead, a patchwork of overlapping and incomplete rights to rule emerges, inextricably superimposed and tangled, in which different de facto juridical instances are geographically interwoven and plural allegiances, asymmetrical suzerainties, and enclaves abound."[44] Mbembe uses Africa as an example of this age of global mobility, and he implies, that because biopower is too closely tied to the logic of state power, it cannot fully make sense of this new global order. Unlike biopower, for Mbembe, necropower can account for the role that killing and death play in global warfare because necropower considers the mobile nature of non- state military forces. For Mbembe, necropower can also recognize that the right to kill does not apply to a sovereign that is subject to its laws, but instead to forms of sovereignty that produce permanent exceptions to the laws.

In addition, Mbembe positions the emergence of an unprecedented form of governmentality within a new geography of the age of global mobility: the management of the multitudes. Population, as a political category in relation to the management of the multitudes, is different from population as a subject category of biopolitical governmentality. Foucault writes that he partly understands governmentality as "the ensemble formed by institutions, procedures, analyses and reflections, calculations, and tactics that allow the exercise of this very specific, albeit very complex, power that has the population as its target, political economy as its major form of knowledge, and apparatuses of security as its essential technical instrument."[45] For Foucault, population as a political category of biopolitical governmentality does not function outside of the state; it is intimately tied to the developments of liberalism, including late liberalism and neoliberalism. For Mbembe, the population in the governmentality of the multitudes is not biopower's target since it is concerned with making a population live. Unlike biopolitics, Mbembe argues, necropolitics can show that the political economy of some governmentalities is "less concerned with inscribing bodies within disciplinary apparatuses as inscribing them, when the time comes, within the order of the maximal economy now represented as the 'massacre.'"[46] There is a generalization of insecurity. The spaces of governmentality of the multitudes are camps and zones of exception. However, I would argue again here that while making live is biopower's primary aim, this does not occlude the necessity for killing within biopower in order to ensure that certain populations do live. Furthermore, the political economy that Mbembe describes in terms of necropolitics is very similar to the political economy of neoliberal biopolitics, since it is concerned with inscribing bodies within a maximal economy of competition.

The Spatio-Temporality of Neoliberalism: Povinelli, Foucault, Cacho, and Wacquant

In this final section, I turn to the spatio-temporality of neoliberalism to show that Mbembe's necropolitics is limited in what it can tell us about the nature of death and suffering under conditions of competition. Again, my goal here is to disentangle Foucault's biopolitics from analyses of the politics of death, such as necropolitics and thanatopolitics, that focus on "outright extinction" in order to highlight the ways that biopolitics do capture ordinary suffering and seemingly ordinary dying. In her recent work, Elizabeth Povinelli has problematized studies that have placed biopolitics' focus on state-theory and on "biopolitically engineered elimination."[47] For Povinelli, this focus on state engineered killing misses the modes of living that exist somewhere in between life and death (a life of struggling to exist).[48] While I agree with Povinelli's criticism of the zero point focus of many studies in biopolitics, I do not think that this means that biopolitics cannot account for ordinary suffering or that biopolitics cannot intersect with state-theory. Seemingly ordinary suffering (whether from chronic illness, depression, drug addiction, or violence) is very much a target of governmentalities that are concerned with making some populations live, since part of the focus of biopolitical systems is to calculate and regulate mortality rates and maximize human capital. Thus, an analysis of neoliberal governmentalities and of some of their agents (including some state agents/agencies) can highlight some of the populations/individuals that these governmental rationalities value. In other words, sometimes, governmentalities as well as states do not kill populations through spectacular modes of killing or other overt forms of killing. Rather, "letting die" is key to contemporary governmentalities, and it can involve neglect, willful ignorance, responsibilization, the romanticization of resilience and burnout, or individualized risk-assessment. All of this produces a spatio-temporality whereby populations of individuals struggle to exist and often are held responsible for their own suffering.

In *Economies of Abandonment*, Povinelli comments on Foucault's and Mbembe's critical theory of life and death. She writes that both Foucault's and Mbembe's perspectives have:

> focused our attention on large-scale killings. In *"Society Must Be Defended,"* for example, Michel Foucault outlined the complex entanglements of sovereign, disciplinary, and biopolitical forms of power at play in the machinery of Nazi genocide. Achille Mbembe situated this industrialized European savagery in a history of African colonization, where colonists experimented with spectacles of irrational, excessive killing.[49]

But, once again, Mbembe's politics of death does not fully capture the spatio-temporality of neoliberalism because it is primarily concerned with death in excess in late-modern colonial occupation and in African colonization. The scope of Mbembe's necropolitics is too large-scale to recognize forms of generalized insecurity and modalities of ordinary suffering under neoliberalism. While Mbembe does not inscribe the right to kill within the state, he still continually seems to assign killing to a clear actor or agent (often, an agent of power). But, it is often the case within neoliberalism that state killing hides behind neoliberal values and ideologies. Thus, it can be difficult to place blame on a particular state actor or to tackle the systemic causes of necropolitics by tracing them back to a central agent/agency. In this last section, I further address the limitations of Mbembe's necropolitics in relation to Povinelli's, Foucault's, Cacho's, and Wacquant's critical theories of life and death in neoliberalism. As I suggested earlier, part of the goal of this study overall, and the aim of this chapter in particular, is to disentangle biopolitics from necropolitics and its focus on "outright extinction" so that we do not miss the subtler forms of death-making that are so characteristic of neoliberalism. That is, the biopolitical is still very much part of the project of neoliberalism as it, independent of its spectacular modes of killing, is still very much about making populations of individuals live in particular ways.

Povinelli argues that "[n]eoliberalism works by colonizing the field of value—reducing all social values to one market value—exhausting alternative social projects by denying them sustenance."[50] Neoliberalism does more than reduce all social values to one market value. Value, whether social or human, is also determined by a logic of competition. This means that whatever helps individuals compete is good. This is consistent with Foucault's characterization of neoliberalism as extending an economic analysis into the non-economic sphere. Foucault once again writes in *"The Birth of Biopolitics"* that "…neo-liberalism seeks… to extend the rationality of the market, the schemas of analysis it offers and the decision-making criteria it suggests, to domains which are not exclusively or not primarily economic."[51] Thus, human value, social value, political value, and educational value (among other forms of valuation) are all subject to the rationality of the market and the logic of competition. For Povinelli, we can understand the social distribution of death through the ways markets and state actors utilize eventfulness.[52] Unlike the spectacular events that unfold in Mbembe's critique of biopolitics and sovereign power, Povinelli is concerned with the uneventful everydayness of the slow deaths that some people experience. These people have for the most part been deemed less valuable by the market. She refers, for example, to a number of indigenous spaces in Australia, writing that "[i]ndigenous communities are often cruddy, corrosive, and uneventful. An agentless slow death characterizes their mode of lethality. Quiet deaths. Slow deaths. Rotting worlds. The everyday drifts toward death: one more drink, one more sore; a bad cold, bad food; a small pain in the chest. Any claim that these forms of decay matter can be referred back to the general condition of human life."[53] These deaths seem "natural" to other populations looking in, but a neoliberal or late-liberal market value is actually what naturalizes these experiences and locates blame within the indigenous community itself.

Earlier in this chapter, I offered an analysis of Mbembe's definition of sovereignty in the context of Foucault's *"Society Must Be Defended."* I suggested that Mbembe conflates sovereign power with biopower in part because he does not clearly distinguish between sovereign power and how sovereignty functions within a mode of biopower. I want to argue, once again, that biopower does not replace sovereign power. Sovereignty and the sovereign right to kill prove insufficient in thinking about subtle forms of state killing, including generalized insecurity and ordinary suffering. But Foucault's biopower, with its focus on endemics, morbidity, and effects on the environment, can account for "slow death"[54] as a neoliberal mode of state killing. Although slow death "occupies the temporalities of the endemic,"⁵⁵ biopolitical governmentalities would consider these slow deaths natural and would work towards managing and perhaps eliminating these forms of slow death in order to ensure the health of the general population. Not only are slow deaths natural in the context of neoliberal biopolitics, but they are also understood through neoliberal values of individual responsibility and the assumption of risk. A study of the biopolitical can thus try to denaturalize forms of slow death by locating unequal death trajectories, limited mobility, and the dailiness of living in the conditions of competition as a consequence, and perhaps even as a given, of neoliberalism.

Biopolitics considers the forms of social existence that may be unique to neoliberalism in ways that necropolitics cannot. It is in considering forms of social existence that I depart from Mbembe's notion of necropower/ necropolitics to argue that what Mbembe and others call necropolitics/ thanatopolitics is indeed biopolitics. Mbembe argues that necropolitics accounts for the conditions that establish the "status of living dead."⁵⁶ This is the point where necropower can still be useful in further concretizing certain forms of governmentality, shifting forms of statehood, and the spatio-temporality of sovereignty. For example, Mbembe examines shifts in the political economy in Africa, ultimately showing how these transformations involve urban militias, private armies, armies of regional lords, private security firms, and state armies,⁵⁷ and how these actors are trying to legitimize the right to kill and commit violence. Within this system, wealthier states can "lease armies to poor state[s],"⁵⁸ and the war machine has emerged as a coercive, violent, and death dealing force. Mbembe explains that these "death-worlds" are the result of a complex relationship between internal, external, and transnational networks of power. He further contends that it is the presence of capital time that has transformed territories into economic enclaves that have become the "privileged spaces of war and death."⁵⁹ The temporality of these "new geographies of resource extraction"⁶⁰ illustrates the ways these war machines speed up and slow down the mobility of surrounding populations, both in prohibiting movement and in forcing displacement.

As a result of these processes, a new governmentality emerges that is more violent because it utilizes "technologies of destruction"⁶¹ that are "more tactile, more anatomical and sensorial."⁶² Moreover, this new governmentality produces conditions of generalized insecurity. While Mbembe does offer and examine concrete examples of the spatial and temporal nature of generalized insecurity, his understanding of generalized insecurity is connected to state and non-state actors. In other words, Mbembe only articulates the condition of the living dead in the context of colonial occupation and its extensions, which are still dominated by large-scale killing. Ultimately, he offers a method for analyzing space and time in a logic of governing that remains dominated by overt forms of killing.

As I suggested above, Mbembe's focus on the logic of the massacre in colonial spaces, and Foucault's analysis of Nazi death camps, are often held as exemplars of the ways studies of biopolitics and their intersections with state-theory understand death-making under neoliberalism and late-liberalism. Again, I argue that biopolitics has the capacity to consider more than overtly violent forms of killing. Rather than focusing on large scale, militarized state, or sub-state killing that operates through the logic of the massacre, we can turn to critiques of neoliberal governmentalities, such as critiques of systems of valuation, critiques of criminalization of surplus bodies, or critiques of neoliberal subjectivities, to only name a few, in order to highlight the less spectacular and taken-for-granted modalities of death-making (ordinary suffering and seemingly ordinary deaths) within biopolitical systems. These critiques may do a better job of taking into account the ordinary suffering and generalized insecurity that are the modus operandi of neoliberal biopolitics.

Lisa Marie Cacho offers one way of understanding forms of death-making and social existence in the context of neoliberal biopolitics. Her concepts of "ineligible for personhood,"[63] or "rightless, living nonbeings,"[64] or even her own understanding of "social death,"[65] can make sense of some of the social existences within the spatio-temporality of neoliberal biopolitics. Cacho considers the outcomes of discourses and governmentalities that produce and regulate neoliberal values/morality. To be considered human, to be valued as a human, and to be recognized by the legal citizen as human, criminalized populations must earn their human status. Not everyone is encompassed within the assumption of inalienable rights since, if certain groups of people must earn their human status or right to become human, then their rights and humanity are never a given.[66]

Furthermore, in tandem with the neoliberal values of individual responsibility and the assumption of risk, forms of morality are also "legally inflicted."[67] Good moral character is, in part, associated with following the law, which means that, if certain groups of people are permanently criminalized, then they can never be understood as "moral or deserving."[68] For Cacho, "to be criminalized is to be prevented from being law-abiding...people who occupy legally vulnerable and criminalized statuses are not just excluded from justice; criminalized populations and the places where they live form the foundation of the U.S. legal system, imagined to be the reason why a punitive (in)justice system exists."[69] She goes on to write that these criminalized statuses are "always already the object and target of law..."[70] Cacho recognizes a relationship between a person's legal status/personhood and the space they inhabit. This relationship is complex in several ways. First, it points to how criminalization often forces bodies into certain spaces of living death.[71] Second, a person's body and location in spaces like the inner city or the global South signal their ineligibility for personhood.[72] Thus, racism produces spaces where people are made permanently vulnerable by global capitalism and neoliberalism. Racism makes possible a state of rightlessness and the criminalization of people who are unprotected by the law, which is similar to the way racism functions for Foucault in biopolitics (i.e., racism normalizes and justifies the death of unwanted populations). But this racism and its state of rightlessness are positive conditions of biopower too. It is not a state of exception that must be induced to justify killing or insecurity, but it is instead a condition that ensures the healthy living of certain populations (to the detriment of others). Healthy living often equates successful competition and the maximization of human capital. Cacho shows that, working along with the law, neoliberalism produces spaces of social death, where the people who do not matter must struggle to live. Thus, Cacho's social death and her analysis of racism can be useful in thinking about the spatio-temporality of neoliberalism in ways that Mbembe's necropower cannot. Using social death to analyze the value systems of neoliberalism can denaturalize certain spaces and their conditions as just the reality and results of unmotivated, criminal, and undeserving peoples.

Loïc Wacquant adds to these conceptualizations of the ways neoliberal biopolitics eliminates dangerous/threatening populations in his book *Punishing the Poor: The Neoliberal Government of Social Insecurity*. Neoliberalism's pull back of social welfare and its increase of insecure work, such as contract work or flexible work, as well as its support and further development of the punitive system, all work together to produce social insecurity. As Wacquant suggests, the "reinforcement and extension of the punitive apparatus"[73] in inner cities and urban peripheries, along with precarious work and weaker social protection, produces social insecurity. Neoliberal values of individual responsibility also contribute to the social insecurity Wacquant critiques.[74] Social insecurity, primarily the punitive mechanism, functions as a way to address potential threats to the good life of the normative/competitive population. And, I would add here that the social insecurity Wacquant and Cacho theorize is not only applicable to criminalized and poor populations. But, rather as I argued in Chapter 2, even middle-class populations of individuals, and as I will show in Chapter 5 wealthy populations, valorize and live by intensified notions of individual responsibility as seen through the self-monitoring cyborg and the biohacker.

Biopolitics, in Wacquant's analysis, does not function in terms of overt state killing as it does for Mbembe. Rather, through the neoliberal state strategy of penalization and through increased flexible work, a number of undesirable and threatening populations (e.g., segments of a population of individuals that fail to compete) are effectively banished from the good population (e.g., segments of a population of individuals that successfully compete). For example, Wacquant writes that, through penalization, "the urban nomad is labeled as a delinquent (through a municipal ordinance outlawing panhandling or lying down on the sidewalk, for instance) and finds himself treated as such; and he ceases to pertain to homelessness as soon as he is put behind bars."[75] Moreover, Wacquant's social insecurity may be better positioned to make sense of neoliberal spatial restructuring and "urban dislocations."[76] Also, Wacquant's social insecurity is more useful in considering the spatio-temporality of neoliberalism because it can highlight its biopolitical workings. Specifically, it highlights the proliferation of new power/knowledge regimes that can manage populations as determined by the conditions of competition.[77]

Conclusion

In this chapter, I have shown that Mbembe argues that Foucault's biopower cannot account for the permanent states of exception and states of siege that make up the spatial and temporal ordering of contemporary colonial occupation or of the governmentality of the management of the multitudes. Furthermore, for Mbembe, the sovereign right to kill is not bound by its own laws within these colonial spaces. I have also argued that Mbembe's focus on spectacular and excessive forms of killing may not help us much to make sense of the spatio-temporality of neoliberalism, and that perhaps we do not need much of an addendum to Foucault's biopolitics to theorize neoliberal governmentalities. If biopolitics can consider the spatio-temporality of neoliberalism, then it can consider ordinary suffering, generalized insecurity, and other forms of death-making that neoliberalism relies on.

As I have suggested in previous chapters, death and killing are positive conditions of biopolitics. The sovereign does not need to suspend the law in order to eliminate certain populations. In fact, as I write in the Introduction, it can appear as though neoliberal governmentality governs without governing as it spreads its subjectivity. The state does not always do the killing or letting die. Rather, in Chapters 1 and 2, I highlighted the ways in which neoliberal biopolitics can ensure the health of some populations through the production, regulation, and management of neoliberal subjectivities that accept, and, in some instances, celebrate an individualism that assumes risk and responsibility for successfully competing. Competing is equated with a life worth living, often with living life. Once again, if the neoliberal subject is not competing, it is not living. Neoliberal subjects that fail to compete or struggle to compete must endure a quasi-life, not quite living and not quite dead, but certainly closer to a living death than more successful populations of individuals.

This is, of course, not to suggest that valorizing competition is acceptable as long as there is a social safety net. Rather, the aim is to point to the ways the individual, as a neoliberal subject, slowly kills itself (smoking, drug abuse, obesity) under the guise of "choice," happily eliminates versions of itself in order to maximize human capital (the quantified-self regulates itself and other individuals to eradicate fat bodies, as we saw in Chapter 2), and willingly endures severe physiological and psychological stress in order to compete. At the same time, I do not mean to suggest here that "willingly" refers to any necessary form of volition, since it is often the case that the neoliberal subject does not have much choice but to compete even though it often thinks of itself as a free individual. The next chapter will continue to problematize neoliberal and biopolitical configurations by making strange the neoliberal governmentalities that permeate cyberpunk city spaces. It will highlight the interplay between these governmentalities, certain discourses, and nodes of knowledge-power that produce and manage the conditions of competition. Chapter 4 will also bring special attention to the spatio-temporality of neoliberalism, or what I call necro-temporality, as it thinks politically about the neoliberal subjects that navigate these cyberpunk spaces.

Notes

1. Karen Yusoff, "An Interview with Elizabeth Povinelli: Geontopower, Biopolitics and the Anthropocene." By Mathew Coleman and Kathryn Yusoff. *Theory, Culture & Society* 34, no. 2–3 (2017), 169–185.
2. Achille Mbembe. "Necropolitics," *Public Culture* 15, no. 1 (2003): 26–29.
3. Mbembe, "Necropolitics," 16.
4. Yusoff, "An Interview with Elizabeth Povinelli," 169–185.
5. Thomas Lemke, "The Risks of Security: Liberalism, Biopolitics, and Fear," in *The Government of Life: Foucault, Biopolitics, and Neoliberalism*, eds. Vanessa Lemm and Miguel Vatter (New York: Fordham University Press, 2014), 68.
6. "Thanatopolitics" is the term preferred by Agamben. See Agamben, *Homo Sacer*.
7. Elizabeth Povinelli. *Economies of Abandonment: Social Belonging and Endurance in Late Liberalism* (Durham: Duke University Press, 2011), 134.
8. Ibid., 11.
9. Michel Foucault, "*Society Must Be Defended*," 255.
10. Ibid., 240.
11. Michel Foucault, "Right of Death and Power over Life," in *History of Sexuality, Volume I* (New York: Vintage, 1990), 135–136.
12. Ibid., 136.
13. Ibid., 248.
14. Mbembe, "Necropolitics," 12.
15. Ibid., 12.
16. Foucault, "*Society Must Be Defended*," 240.
17. Ibid., 254.

18. Foucault, *History of Sexuality*, 139.
19. Ibid., 139.
20. Ibid., 139.
21. Foucault, *"Society Must Be Defended,"* 240.
22. Mbembe, "Necropolitics," 17.
23. Ibid., 17.
24. Ibid.
25. Foucault, *"Society Must Be Defended,"* 256.
26. Mbembe, "Necropolitics," 16.
27. Ibid., 16.
28. Mbembe, "Necropolitics," 12.
29. Ibid., 18.
30. Ibid., 19.
31. Ibid., 21.
32. Ibid., 23.
33. Foucault, *"Society Must Be Defended,"* 257.
34. Mbembe, "Necropolitics," 24.
35. Giorgio Agamben, "Threshold," in *Homo Sacer: Sovereign Power and Bare Life*, trans. by Daniel Heller-Roazen (Stanford: Stanford University Press, 1998), 181.
36. Foucault, *"Society Must Be Defended,"* 37.
37. Mbembe, "Necropolitics," 23.
38. Foucault, *History of Sexuality*, 137.
39. Jean-Paul Sartre, preface to *The Wretched of the Earth* (New York: Grove Press, 1961), 7–34.
40. Foucault, *"Society Must Be Defended,"* 245.
41. Frantz Fanon. *The Wretched of the Earth*, trans. C. Farrington (New York: Grove Weidenfield, 1991), 37.
42. Mbembe, "Necropolitics," 27.
43. Foucault, *"Society Must Be Defended,"* 49.
44. Mbembe, "Necropolitics," 32.
45. Michel Foucault. *Security, Territory, Population, Lectures at the College De France, 1977–78*, translated by Graham Burchell (New York: Palgrave MacMillan, 2009), 144.
46. Mbembe, "Necropolitics," 34.
47. Elizabeth Povinelli, "An Interview with Elizabeth Povinelli: Geontopower, Biopolitics and the Anthropocene," By Mathew Coleman and Kathryn Yusoff. *Theory, Culture & Society* 34, no. 2–3 (2017): 169–185.
48. Ibid., 170.
49. Elizabeth Povinelli, *Economies of Abandonment*, 134
50. Ibid.
51. Michel Foucault. *The Birth of Biopolitics: Lectures at the Collège de France, 1978–79*, trans. Graham Burchell (New York: Palgrave MacMillan, 2008), 323.
52. Povinelli, *Economies of Abandonment*, 134.
53. Ibid., 145.
54. Elizabeth Povinelli builds upon Lauren Berlant's phrase "slow death." According to Lauren Berlant, slow death is "the physical wearing out of a population and the deterioration of people in that population that is very nearly a defining condition of their experience and historical existence." See Laurent Berlant, "Slow Death (Sovereignty, Obesity, Lateral Agency)," *Critical Inquiry* 33 (2007): 754.
55. Lauren Berlant, "Slow Death," 756.

56. Ibid., 40.
57. Mbembe, "Necropolitics," 32.
58. Ibid., 32.
59. Ibid., 33.
60. Ibid., 34.
61. Ibid., 34.
62. Ibid., 34.
63. Lisa Mari Cacho, *Social Death: Racialized Rightlessness and the Criminalization of the Unprotected* (New York: New York University Press, 2012), 6.
64. Ibid., 6
65. Ibid.
66. In Chapter 1, I showed that cyborganization in cyberpunk can remind us that human status is not a given. The genre strategies of cyberpunk, such as amplification and cyborganization, are not only ways in which neoliberalism constructs and maintains its reality within the narrative realm of cyberpunk science fiction novels, but they also play a generative role in the production of "real" neoliberal values, neoliberal discourses, and neoliberal subjects. Cyberpunk, with its political ambiguity about the liberal humanist tradition, its examination of what kinds of life are proper to a particular political system and economic system, and its dramatization and visualization of the spatio-temporalities of neoliberalism, offers us a kind of biopolitics.
67. Ibid., 4.
68. Ibid., 4.
69. Ibid., 5.
70. Ibid., 5.
71. In Chapter 1, I showed that cyberpunk dramatizes and visualizes this complex relationship between an individual's human status and the spaces these individuals live in (e.g. criminalized replicants and their movement through insecure and dangerous parts of Los Angeles).
72. Ibid., 6.
73. Loïc Wacquant. *Punishing the Poor: The Neoliberal Government of Social Insecurity* (Durham: Duke University Press, 2009), 5.
74. Ibid., 5.
75. Ibid., xxi.
76. Ibid., xxi.
77. Ibid., 31.

Bibliography

Agamben, Giorgio. "Threshold," in *Homo Sacer: Sovereign Power and Bare Life.* translated by Daniel Heller-Roazen. Stanford: Stanford University Press, 1998.

Berlant, Lauren. "Slow Death (Sovereignty, Obesity, Lateral Agency)." *Critical Inquiry* 33 (2007): 754–780.

Cacho, Lisa Marie. *Social Death: Racialized Rightlessness and the Criminalization of the Unprotected.* New York: New York University Press, 2012.

Fanon, Frantz. *The Wretched of the Earth.* translated by C. Farrington. New York: Grove Weidenfield, 1991.

Foucault, Michel. "Right of Death and Power over Life," in *The History of Sexuality: An Introduction, Volume I.* New York: Vintage Books, 1990.

—., "Lecture 11," in *Society Must Be Defended: Lectures at the College De France, 1975–76.* translated by Macey, David, New York: Picador, 1997.

—., "Lecture 8-Lecture 12," in *The Birth of Biopolitics: Lectures at the Collège de France, 1978–79*, translated by Burchell, Graham, New York: Palgrave MacMillan, 2008).

—., "Security, Territory, Population," *Lectures at the College De France, 1977–78*, translated by Burchell, Graham, New York: Palgrave MacMillan, 2009.

Lemke, Thomas. "The Risks of Security: Liberalism, Biopolitics, and Fear," in *The Government of Life: Foucault, Biopolitics, and Neoliberalism*, edited by Vanessa Lemm and Miguel Vatter, New York: Fordham University Press, 2014: 59–76.

Mbembe, Achille, "Necropolitics," *Public Culture* 15, no. 1 (2003): 11–40.

Povinelli, Elizabeth A. *Economies of Abandonment: Social Belonging and Endurance in Late Liberalism*, Durham: Duke University Press, 2011.

—., "An Interview with Elizabeth Povinelli: Geontopower, Biopolitics and the Anthropocene," by Mathew Coleman and Kathryn Yusoff. *Theory, Culture & Society* 34, no. 2–3 (2017), 169–185.

Sartre, Jean-Paul. *Introduction to the Wretched of the Earth*, 7–34. New York: Grove Press, 1961.

Wacquant, Loïc. *Punishing the Poor: The Neoliberal Government of Social Insecurity*, Durham: Duke University Press, 2009.

4 Cyberpunk Necroscapes and Necro-temporality in *Blade Runner*

Introduction

In many ways, this chapter is a continuation of the argument developed in Chapter 1. Among other things, this chapter builds upon this project's attempts to move beyond representation as a method for analyzing literature and film. Again, in this project, I do not wish to offer readings of films, in this case cyberpunk films, that act simply as evidence for the existence of a postmodern condition, late-capitalism, or even the forms of competition that I suggest are very much a reality of neoliberalism. I am not trying to diagnose our current conditions, nor do I want to offer readings of cyberpunk films that reaffirm this diagnosis. Instead of focusing on cyberpunk novels as I did in Chapter 1, in this chapter, I perform a kind of "cognitive estrangement" with structures and assumptions, namely with the spatial and temporal conditions of competition and the assumption of risk, that undergird, as Steven Shaviro suggests, our social reality, particularly as these structures, assumptions, and social realities are performed by film, in this case, the cyberpunk film *Blade Runner*. Like Shaviro, I accept that films "...are also productive, in the sense that they do not *represent* social processes, so much as they participate actively in these processes, and help to constitute them."[1] But, rather than call this form of defamiliarization a mapping of affective and information flows, as Shaviro does in his later work, *Post-Cinematic Affect*, I stick with his earlier use of Carl Freedman's term "cognitive estrangement," a term that Freedman himself borrows from Darko Suvin.[2] "Mapping the flows of affect" often applies to spaces that are occupied by the populations that compete in ways that are more intelligible, and thus, more valued by neoliberal governmentalities that are aimed at nurturing and inciting competition. I suggest in this chapter that "cognitive estrangement" can make strange conventionally held notions about how we ought to live our life and make intelligible less valued populations as they compete to live in the spatio-temporality of neoliberalism.[3]

Shaviro's "Mapping the flows of affect" is too similar to Jameson's "cognitive mapping," in the sense that both assume that within the spaces of the present (Shaviro borrows Augé's concept of non-place and, for Jameson, it is postmodern hyperspace) the individual has lost the ability to locate "itself, to organize its immediate surroundings perceptually, and cognitively to map its position in a mappable external world."[4] Thus, the goal for Jameson, and it seems for Shaviro in *Post-Cinematic Affect* too, is to produce an "aesthetic of cognitive mapping" whereby the individual subject can gain "a heightened sense of its place in the global system..." and "...begin to grasp our positioning as individual subjects..."[5] I agree that maps are not just representations of space, and that maps are "tools for negotiating, and intervening in space."[6] But the act of mapping is also caught up in a history of imperial and colonial domination.[7] Jameson and Shaviro seem to accept the act of mapping uncritically. There is no consideration for who has traditionally done the mapping over space and people. The goal in performing this cognitive estrangement or defamiliarization with the spatio-temporality of neoliberalism, in part, is to make the boundary between science fiction and our current conditions (Donna Haraway suggests that this boundary is an "optical illusion"[8]) less distinct, less concrete, and less taken-for-granted. By making this boundary less concrete, I hope to, again, as I tried to suggest in Chapter 1, make fictive/strange the social/economic status-quo, and thus open the possibility for seeing beyond the conditions of competition.

I associate mapping the flows of affect and information with privileged populations because I think that this mapping captures the movement of populations that have human capital, especially those populations of individuals that have the resources to live intensely and to continually maximize their human and social capital. Shaviro writes that the spaces of neoliberalism elicit a sense of transience. Borrowing from Marc Augé's notion of non-places and Gilles Deleuze's any-space-whatevers, Shaviro characterizes these spaces as sleek and anonymous.[9] According to Augé, non-places are "a world where people are born in the clinic and die in hospital, where transit points and temporary abodes are proliferating under luxurious or inhuman conditions...where a dense network of means of transport which are also inhabited spaces is developing; where the habitué of supermarkets, slot machines and credit cards communicates wordlessly, through gestures, with an abstract, unmediated commerce; a world thus surrendered to solitary individuality, to the fleeting, the temporary and ephemeral."[10] I agree that the non-places/any-space-whatevers/sleek, anonymous spaces that Augé, Deleuze, and Shaviro consider are part of the spaces that emerge as a result of the conditions of competition and the society of control that thrive in the context of neoliberalism and its governmentalities. But I do not think that these notions fully capture the ways some populations without human capital or with less human capital must encounter and move through these non-places or any-space-whatevers.

There is a problematic assumption here that individuals without human capital or with less human capital can actually move through these non-places, including hotels, clubs, gyms, supermarkets, malls, and suburbs, to only name a few, and thus acquire the credit cards they need to move as swiftly as the information that flows through these spaces. Moreover, these spatial theories do not fully capture the spatio-temporalities of neoliberalism and its governmentalities. While it is true that these populations with no or less human capital often surrender to the temporary/transient nature of the non-place, it is not for the same reasons as the populations with human capital, since the suggestion is that more competitive populations move quickly through and stay in these non-places temporarily because they now move, think, and live like the information that flows through these networked spaces, and not because their living in/occupation of these spaces is illegal, unwanted, or eliminated.

Shaviro suggests that these non-places often play a bigger role in some films than the bodies that navigate these spaces and its flows of affect. He focuses specifically on the film *Boarding Gate*.[11] However, his point about the more "active role" that non-places play in some films is applicable to most cyberpunk films, since it is often the case that these films, including *Blade Runner*, take place within an all-encompassing mega-city and its ever-expanding uniformity. Often, a problem with analyses of cyberpunk films, especially in the case of *Blade Runner*, is that they treat these non-places almost as though they are characters in a film, and the populations that move through these non-places at times become the background for or secondary accessories to these non-places. The fear in these analyses, which often borrow from Jameson's critique of the cultural logic of late-capitalism, is that time and the self have become subordinate to space. This becomes a space that is, on the one hand non-representable, and yet, on the other hand, according to these analyses (I am thinking in particular of Harvey's and Bruno's respective analyses of *Blade Runner*, since they are especially influential on the way later critiques looked at cyberpunk films), film can aptly represent this postmodern spatial logic, that is both the non-place and part of the logic of late-capitalism. Cyberpunk films are then no more than a mirror image of the real. Thus, what the cyberpunk film might do (e.g., actively produce and participate in neoliberal governmentalities and its subjectivities, make strange the conditions of competition and neoliberal values, and/or frame the ways in which hyperobjects are made intelligible), sometimes in spite of itself, is ignored or foreclosed.

In this chapter, I work against the time of the non-place/any-space-whatever, the network society, the society of control, and schizophrenic temporality that Augé, Manuel Castells, Deleuze, Shaviro, and Jameson, among others, characterize as time-less time. The suggestion again and again is that the "spatial logic of the database" or the spatial logic of the digital/virtual has replaced that of narrative (be it literature or film time), cyclical time, or clock-time.[12] Or, in the case of Deleuze's "Postscript on the Society of Control," the society of control has replaced the society of discipline, and while Deleuze does not use the term "time-less time," it seems to be applicable to the flexible, transient, rapid, and free-floating spaces of the society of control. But, I am not convinced that these attempts at making a hyperobject (e.g., massively distributed financial/digital networks) intelligible or at diagnosing the space and time of the present fully appreciate the spatio-temporality of the conditions of competition and its governmentalities for several reasons.

First, time-less time suggests that duration, including the duration of a person's life, how long someone works, how long someone exercises, how long a person occupies a certain space, and so on, is no longer possible or important. There still are other political/social/economic/ethical systems and practices that exist along with and that even further neoliberalism and that still encourage populations of individuals to spend a certain amount of time working, living, sleeping, exercising, etc. Even if thirty minutes of exercise, eight hours of work, or twelve hours of living can all fall under 24/7 labor time[13] (it is all work time geared towards maximizing human capital), duration and calculability are still very much a part of the ways individuals and governmentalities produce and manage subjectivities. The nodes of power-knowledge, including neoliberal health and the medical industry, and even governmentalities that cultivate the conditions of competition (e.g., marketization of space, the assumption of risk, the intensification of the responsibilized-self, and the logic of intensity) continue to play a role in the production, management, and control of space. As we saw in Chapter 2, duration and the quantification of time are both a part of these governmentalities and nodes of power-knowledge, since a good neoliberal subject, which includes the self-monitoring cyborg, must have ways to calculate the degree to which they have maximized their human capital.

Second, time-less time points to the loss of "the notion of the lifecycle,"[14] which I would argue misses the fact that neoliberalism and biopolitics reinforce each other. The society of control has not necessarily replaced the society of discipline. "Replace" may be too strong a term here. Unlike Deleuze and Shaviro, I suggest that the society of control exists along with discipline, biopower, sovereignty, etc. The emergence of one form of power does not mean the end of an older form of power. That is, biopower does not replace sovereign power or disciplinary power. As Foucault suggests, the law (sovereign power) does not disappear or fade away. Rather it is shaped, produced, and managed by other forms of power that exist along with it, including biopower, disciplinary power, and control.[15] Thus, the spatio-temporality of the society of control, I argue, is still shaped by neoliberal biopolitics, its governmentalities, and nodes of power-knowledge, including health discourses, the privatization of the penal system (which, according to Deleuze, in the society of control, is on its way out), the fitness industry, predictive technologies, and so on. Biopolitics is still very much a part of the present. It is not, as Shaviro suggests, that Foucault tells us to "follow the proliferation of market logic"[16] instead of biopolitics/biopower, as though the proliferation of market logic were antithetical to biopolitics or that it would replace biopolitics. Rather, the logic of the market is, in part, how biopower/biopolitics functions. It is through economic rationality that biopolitics calculates aleatory events, manages life, and defines the forms of life that are subject to, intelligible for, and subjectivized by neoliberalism. Neoliberalism and biopolitics both thrive under and through an economization of the social/political/biological, and they accommodate and perpetuate each other.

And, third, the part of timeless-time that denies mortality, as Shaviro and Castells suggest time-less time does, once again, might account for the ways some privileged populations move through and live in nonplaces and other neoliberal spaces. That is, the logic of intensity, transhumanism, and biohacking, to only name a few, reflect a general disdain for death as a failure to compete successfully, and thus, to live a good life. But a denial of mortality may only account for the ways some populations of individuals subjectivize themselves as they encounter the spatiotemporality of neoliberalism and its conditions of competition. From the beginning of this study, I have suggested that the spaces and temporalities of neoliberalism are necrotic. People are generally more insecure as they intensify their acceptance of risk and responsibility for maximizing their human capital and avoiding burn out, but clearly, some populations have less resources, less money, less/no capital, and thus struggle or fail to compete within the marketized spaces of the present. They then must endure a quasi-life since competition is equated with living.

In this chapter, I elaborate on what I consider to be the necrotic time of neoliberalism, or what I call necro-temporality, as a way to work against the notion of time-less time, and as part of the cognitive-estrangement I will perform on the cyberpunk film, *Blade Runner*. I perform a cognitive estrangement from the dominant neoliberal notions about how we ought to live our lives that *Blade Runner* normalizes as I disentangle *Blade Runner* from Jameson's cognitive mapping and from Bruno's postmodern aesthetics. For many cultural critics, the city of cyberpunk is thoroughly postmodern. These critics draw from Jameson's essays "Postmodernism, or the Cultural Logic of Late Capitalism" and "Postmodernism and Consumer Society" to argue that, like the cyberpunk city, the postmodern condition is characterized by a spatial pastiche, schizophrenic temporality, fragmented subjectivity, simulation, and a waning of affect. Jameson's postmodern aesthetics can be traced through the majority of the scholarship on *Blade Runner*. For example, Bruno's essay

"Ramble City: Postmodernism and Blade Runner" is an early and often discussed work that applies Jameson's postmodern aesthetics to *Blade Runner*.

Bruno writes that *Blade Runner* is a "metaphor of the postmodern condition."[17] As I pointed out in the Introduction, Bruno analyzes the architectural aesthetics of the film through Jameson's understanding of the postmodern condition. She expands upon Jameson's notion of spatial pastiche, suggesting as Jameson does, that pastiche holds a privileged position in studying postmodern aesthetics. However, Bruno's focus on the film's architecture ignores the micro-political encounters that occur within the city. In other words, bodies disappear in Bruno's analysis of postmodern aesthetics. Rather than examine the systemic causes of the necrotic material conditions of Scott's city, she views them as aesthetic markers of postmodernism.

According to Jameson, contemporary theory destabilizes the subject by questioning its legitimacy. The destabilized or fragmented subject is not an autonomous individual, and thus poses a problem for expression, since styles are dependent on the notion of an "individual monad."[18] Postmodernism replaces the modern subject with a new one that signals the end of "...the bourgeois ego or monad...of style, in the sense of the unique and the personal," and it liberates the postmodern subject from "expression and feelings or emotions, not merely a liberation from anxiety, but a liberation from every other kind of feeling as well, *since there is no longer a self present to do the feeling.*"[19] Within the context of Jameson's postindustrial aesthetics, *Blade Runner* explores this postmodern subjectivity, in part through its representation and examination of memory, emotions, and self-hood, suggesting that the modern subject, with its ability to feel and remember, defines what it means to be human, as opposed to the postmodern subject. Bruno's analysis reflects a thread in film theory, including Jameson's "Postmodernism, or the Cultural Logic of Late Capitalism" and "Postmodernism and Consumer Society" and David Harvey's *The Condition of Postmodernity*, that often ignores the ways neoliberal governmentalities construct ontological categories in relation to urban spaces and the ways in which individual subjects shape their subjectivity in relation to these governmentalities.

Instead of reading *Blade Runner* within a postmodern aesthetics or as a mirror image of the real, I depart from a postmodern aesthetic analysis of the replicants' sense of time, relation to space, fragmented subjectivity, and a waning affectivity, to argue instead that what seem like markers of postmodern aesthetics (e.g., schizophrenic temporality and spatial pastiche), of time-less time, or of non-places, are instead the ways in which Scott's film perpetuates a logic of intensity. And, I suggest that this logic of intensity, as a condition of neoliberalism, is part of what produces necro-temporality. As I indicated above, I work against the notion that the time of the non-place, the any-space-whatevers, the network society, and schizophrenic temporality fully captures the time of neoliberalism.

I will do this, in part, by offering a cognitive estrangement of the spatio-temporalities of the conditions of competition and the logic of intensity in *Blade Runner*. Through this cognitive estrangement, what *Blade Runner* does as an active producer and participant in neoliberal governmentalities and its subjectivities will be brought to light. This, of course, is not to suggest that Jameson, Bruno, and Harvey, among others, are wrong in their reading of Scott's *Blade Runner* as a mirror image of the real or as a reflection or confirmation of postmodern conditions. Rather, my point, is that films, in this case *Blade Runner*, can do more than confirm theoretical diagnoses of postmodernism or other systemic problems.

I want to suggest here, that despite its ambivalent politics, *Blade Runner* actively produces and participates in neoliberal governmentalities by contributing to the normalization of individualized risk assumption and responsibility. The world of *Blade Runner* is one where populations of individuals cannot and should not rely on systemic support. Replicants along with other humans must rely on themselves to survive and endure the depravations of neoliberal capitalism. The Tyrell Corporation incites replicants and humans to live intensely, to embrace the dangers of neoliberalism in order to live brightly (to burn brightly), to live on the edge of burn out, in a world that has already burned out. Deckard has no idea if Rachel will live past a four-year life span, but this insecurity only makes his life more intense, only incites him to embrace every moment of life as though he and Rachel may die tomorrow. And, of course, neoliberalism benefits from this intensification of living, an intensification of "seize the day," since the neoliberal subjects who "seize the day" will not necessarily see the depravations of neoliberalism and its capitalism as a problem. That is, these depravations, whether it is environmental destruction, increased poverty, isolation, abandonment, Social Darwinism, etc., become challenges to proudly overcome, and overcoming becomes an indicator of a person's value, character, and resilience. *Blade Runner* thus appeals to and perpetuates notions of individualism, freedom, and risk that are consistent with neoliberalism. It does not simply reflect the cultural and political conditions of its time. *Blade Runner* reminds individuals that the corporate state provides no safety net and that the state ought not to provide a safety net because it is corrupt and inefficient and because it limits freedoms. Instead, *Blade Runner* incites individuals to "seize the day," to accept the risks of life, to take care of themselves because no one else will (the state certainly will not). The point for the film *Blade Runner* is not that the political and economic systems are horrific, but rather that even in the face of neoliberal depravations there is hope (e.g., the human can maintain some semblance of its humanity and there is beauty even in hell). In other words, *Blade Runner* tells us that we must remain resilient.

Cognitive estrangement then is about the ways in which we think about film. It is thinking politically about films in a way that recognizes how they actively produce and participate in the cultural, economic, social, and political systems within which they exist. In this case, I am thinking politically about *Blade Runner* in a way that moves against the grain, that is to say, that makes the notion of living life to the fullest, of seizing the day, strange by suggesting that this attitude towards how we live our life reflects a kind of necro-temporality. Making strange here is another way to make neoliberalism unfamiliar. In other words, I am making strange "living life to the fullest" and "seizing the day," advice to live life that are traditionally thought of as admirable, good, a sign of individuality, etc., by suggesting that these notions are in fact part of a logic of intensity that is very much a condition of neoliberalism. Wanting to live life intensely is imminently governable for neoliberal governmentalities. *Blade Runner* offers a visual refashioning of neoliberal governmentalities that points to the ways these governmentalities cultivate and "make real" competition, the logic of intensity, and risk. Thus, thinking about cyberpunk cities in terms of necro-temporality can "generate ways to think 'the political'"[20] that remain aware of the necrotic conditions that many bodies face in cityscapes.

A Brief Return to Mbembe: Necroscapes and Necro-temporality

Achille Mbembe defines necropower as the modern state's ability to reduce human life to its bare or instrumental forms in order to materially destroy human bodies. Necropolitics is the subjugation of life to the power of death. Mbembe is largely concerned with the ways "weapons are deployed in the interest of maximum destruction of persons and the creation of *death-worlds*."[21] I extend this interest in the destruction of persons and creation of death worlds into an analysis of urban design. And, as I argued in Chapter 3, the destruction of persons and creation of death worlds is not antithetical to biopolitics. Killing and letting die are both positive conditions of biopower. Biopower kills in order to make some populations live. It may not always be so clear that a state is doing the killing. For Mbembe, necroscapes are spaces (sometimes cities) that are designed to capitalize on this reduction of human life to its bare forms by destroying unwanted bodies. Through urban design, securitization, terror, state violence, surveillance, and the production of vulnerable bodies, necroscapes kill certain bodies. Necroscapes let bodies die by restricting access to health care, food, water and other resources. These cities use these subtle and overt ways of killing unwanted bodies. For the purpose of this chapter, I amend Mbembe's conception of the necroscape to suggest, instead, that necroscapes are not only cities but also spaces over which the conditions of competition (including the marketization of space), the logic of intensity, and risk have spread.

Mbembe is not explicit about the nature of time in necroscapes. He writes about different temporalities in necroscapes that produce life as "a form of death-in-life." Borrowing from Giorgio Agamben's *Homo Sacer*[22] and Carl Schmitt's state of exception,[23] Mbembe understands the modern state's ability to deny political subjectivity, reduce life to bare life, and exclude these bodies from the "normal state of law" as setting the conditions for and in necroscapes. He sees plantation temporality, apartheid temporality, and colonial temporality as three key manifestations of the state of exception and as precursors to necroscapes. His understanding of certain bodies in necroscapes is that they are a kind of living dead; these bodies are from birth subject to the state's power to let die or kill.

I use necro-temporality as a way to describe and problematize the temporalities and spatialities of neoliberalism and its governmentalities. I am especially concerned with the spatio-temporalities of the conditions of competition, which I consider to be a key dimension of neoliberalism, as an example of necro-temporality. As I argued in the Introduction to this study, nodes of knowledge-power work in tandem with the law, the logic of competition, and often a new military urbanism to produce and control necroscapes and necro-temporalities. Within necroscapes, risk and responsibility, among other effects, are intensified, and the populations of individuals that move through and live in these necroscapes are always potential threats to market competition. Necroscapes are marketized spaces where populations of individuals become at once targets of and threats to competition. If these populations of individuals want to move and live freely within necroscapes, they must prove their competitive worth.[24]

I understand necro-temporality as a kind of time produced by neoliberalism. In Chapter 3, I made the case that killing and letting die certain populations are not conditional upon suspending the law or reducing individuals to bare life. This is true for necro-temporality as well; it is not dependent upon the suspension of law, since again, the death of some populations, whether through spectacular state killing or more subtle forms of letting die, is a positive condition of biopolitics. Under neoliberalism, necro-temporality is a precarious state of existing in time whereby individuals are insecure and the responsibility they have for their own insecurity is intensified. Furthermore, necro-temporality is also about certain forms of death/dying caused by stress and insecurity, and yet these deaths are understood as a result of weak neoliberal character/values and/or as a failure to compete, and thus, these deaths are just a "natural" part of the human condition. Necro-temporality, then, is a biopolitical technique for neoliberal governmentalities to regulate and manage death (how it is defined, treated, and dealt with). It makes it easier to dismiss or ignore the death of some populations as natural, inevitable, unsolvable, or as a failure in character. Necro-temporality points to the ways power makes some populations vulnerable to death and exploitation by limiting time in certain spaces, by forcing continual mobility, increasing time in toxic, violent, and surveilled spaces, and by justifying these necro-temporal policies through racism, neoliberal morality, the logic of intensity, and the naturalization of the free market and competition.

Schizophrenic Temporality and Blade Runner

Jameson and Bruno draw their insights from a Lacanian understanding of temporality. Lacan argues that temporality is connected to a linguistic order whereby language constructs an understanding of past, present, future, and memory.[25] In order to have a history or be a part of history, subjects must be able to participate within a linguistic order that allows them to construct a linear narrative of time. Bruno writes: "...the experience of temporality and its representation are an effect of language. It is the very structure of language that allows us to know temporality as we do and to represent it as a linear development...[H]istorical continuity is...dependent upon language acquisition."[26] The suggestion here is that the replicants, genetically engineered humans in *Blade Runner*, must enter the dominant symbolic order to have a past, present, and future, and because they fail to enter this symbolic order, they represent a "new form of temporality, that of schizophrenic vertigo."[27] The failure on their part to integrate into the symbolic order represents the fragmentation of subjectivity in postmodern reality. For Jameson and Bruno, replicants are incapable of historical continuity because they do not have a centered subjectivity that works within a linguistic order. Thus, they are stuck within a perpetual present, or in what Jameson calls "a series of pure and unrelated presents in time."[28] Jameson makes the case that identity "...is itself the effect of a certain temporal unification of past and future with the present before me."[29] In the context of Jameson's postmodern condition, the replicants, then, do not have a personal identity, at least not one that the state recognizes or even values, nor do they have subjectivity. The power to unify the past and future with a present allows some populations to write history and exclude other bodies from this history.

Yet, it is unclear exactly how the replicants fail to enter the dominant symbolic order. Bruno characterizes the replicants as simulacra and suggests that their simulative condition leads to their inability to participate in the dominant symbolic order. I question whether, as Bruno argues, the "narrative 'invention' of the replicants is almost a literalization of Baudrillard's theory of postmodernism as the age of simulacra and simulation."[30] It is not that the replicants simulate human emotions, sexuality, memory, or pain. Rather, as genetically enhanced humans, they have developed emotions, sexualities, and memories as a different race or species. At the end of his life, Roy Batty, a "replicant," meditates on his existence when he tells Rick Deckard, a blade runner, that "I've seen things you people wouldn't believe. Attacked ships on fire on the shoulders of Orion. I watched C-beams glitter in the dark near the Tannhauser gate. All those moments will be lost in time...like tears in the rain...Time to die."[31] Batty recognizes that replicants are different; this difference is clear when he references humans as "you people," differentiating replicants' experience of pain, enslavement, and beauty from that of humans. Replicants may look human, but their perfect forms, increased strength, high endurance, and reflective eyes are clear indicators of their more than human status. He locates his life within a grand narrative that clearly has a rich understanding of and place in time; even though, he recognizes that once he dies his life will be forgotten by humans. Their use of romantic poetry and discussions about complex biomechanics, along with a clear sense of their history, suggest that it is not a failure on their part to enter the symbolic order.

Bruno writes that "[r]eplicants are condemned to a life composed only of a present tense: they have neither past nor memory."[32] Yet Roy Batty shares some of his memories with Deckard, indicating that he has a past, that he remembers his past, and that it is the collective replicant past that has driven them back to earth to seek an extended life. Bruno goes on to write: "[t]heir assurance of a future relies on the possibility of acquiring a past. In their attempt at establishing a temporally persistent identity, the replicants search for their origins. They want to know who 'conceived' them, and they investigate their identity and the link to their makers."[33] Roy and his cohort return to earth in order to acquire a future. They know that the Tyrell Corporation engineered their bodies so that they would live for only four years. Replicants endure horrific violence and likely death in order to demand extended life from their makers. Their future is not dependent on acquiring a past as they already have one; rather, the future for replicants is unsure because of the neoliberal biopolitics of the Tyrell Corporation. To suggest they have no past, as Bruno does, and the Tyrell Corporation and blade runners do, ignores their enslavement and perpetuates a temporality that justifies their death: a necro-temporality.

Bruno's analysis of *Blade Runner* places the film firmly within a postmodern aesthetics. She later writes: "[r]eplicants represent themselves as a candle that burns faster but brighter and claim to have seen more things with their eyes...than anybody else."[34] Instead, it is Eldon Tyrell, their maker, who characterizes replicants as brightly burning candles, when he tells Batty, "the light that burns half as long, burns twice as bright... You're quite a prize...Revel in your time."[35] This exclamation justifies a logic of intensity: the implication here is that Roy should appreciate the fact that he lived the epitome of an intense life. If the replicants are simulacra, they are in fact simulacra of the perfect neoliberal individual that lives intensely to the point of burn out and then has the good grace to eliminate its surplus body from the population. Of course, Roy Batty and his cohort problematize their role as the perfect neoliberal individual when they resist having the good grace to accept their early death (their elimination from the population when they reach burn out). Replicants do not see themselves as short lived yet intensely alive beings. They wish to extend their life rather than accept their fate. Humans (or the neoliberal order), on the other hand, produce replicant temporality (neoliberal necro-temporality) as short and intense in order to justify their enslavement, murder, and morphology. In fact, the term "replicant" already inscribes their being within a temporal narrative that reaffirms the superiority of humans (or neoliberalism) as the authentic and dominant paradigm. Again, as I suggested above, the point is not necessarily that Bruno is wrong, but that *Blade Runner* can do more than passively reflect the postmodern conditions of its time and reaffirm Jameson's and Bruno's reading. *Blade Runner* is much more productive than what Bruno's theory of representation suggests.

Necro-temporality and Blade Runner

The ways power forces in *Blade Runner* exclude replicants from history, limit time, force continual movement, and increase time in toxic, violent, and surveilled areas are all examples of necro-temporality. Bruno's analysis misses what Mbembe makes clear: in the context of the system that propagates their enslavement (off-world colonies), "the humanity of the slave [replicant] appears as the perfect figure of a shadow."[36] Replicants are shadows, not because of an inability to enter a symbolic order, but because of their vulnerability as slaves, runaways, and rebels. As slaves in off world colonies, replicants do not exist outside of their role as property. Their labor plays a part in aiding (at least some) humans minimize the time they spend in hard labor. The Tyrell Corporation treats replicants as though they do not exist, "except as a mere tool" in order to maintain their status as free labor. The corporate state in *Blade Runner* produces a racist narrative that places replicants into a less than human subgroup. This narrative understands replicants as incapable of memory, emotions, and identity, thus making them unable to participate in history. Their less than human status combined with their exclusion from human history makes it easier for the corporate state to regulate how replicants experience time. Time for replicants is regulated by the dominant race/corporate state.

One way the dominant race regulates time is through laws that restrict how much time replicants spend on earth. During the opening credits, the audience is informed that, after a bloody mutiny, replicants were "declared illegal on earth." Returning to earth is punishable by death, which is known as "retirement," not execution. Thus, replicants are not allowed to spend any time on earth. This law is an example of the extreme anti-immigration policies often deployed by neoliberal governmentalities. By making the return to earth punishable by death, earth becomes a place where security forces can hunt and kill any illegal immigrants (replicants). No one is willing to help them because security forces and the Tyrell Corporation use racism to convince the public that replicants are unfeeling, dangerous, and predatory. And, the marketized spaces that naturalize competition ensure that replicants (populations without or with little human capital) are blamed for their failure to compete. Even if a replicant could manage to evade a blade runner, it would not live past a four-year life span, as Roy exemplifies. There would never be too many illegal replicants to compete for jobs and resources.

Security forces, in part, understand the social order as divided between the little people and those who kill. The police chief exemplifies this power over life when he says to Deckard, "[y]ou know the score pal, if you're not a cop you're little people."[37] The implication here is that if you are not a cop, you are powerless. Later, Deckard reflects on his condition as a soon to be out of retirement blade runner, "I'd quit because I'd had a belly full of killing...But I'd rather be a killer than a victim."[38] Thus, time and space are divided between the people who seemingly have power and the people who are subject to necro-temporality. I use "seemingly" as a reminder that all populations are subject to necro-temporality under neoliberalism, but necro-temporality remains differentially experienced.

Their illegal status limits the amount of time replicants can spend in one place. They must remain mobile in order to avoid being killed by blade runners. Replicants are not the only populations of individuals that are exploited, hunted, or made insecure within the spaces of *Blade Runner*. But replicants' limited life span forces them to move quickly through the levels of the city in order to find a way to live longer. Replicants, like the other unwanted populations of *Blade Runner*, hide in the shadows, work in the bowels of the Tyrell Corporation, endure exploitative conditions, and temporarily inhabit spaces for sleep and sustenance. The constant movement from place to place on earth makes replicants an invisible population; no one sees them because they hide and move in darkness. But while the film in some ways does highlight the problems of illegal immigration, the illegal immigrants remain white. Thus, I would suggest that the film also reinforces the notion that we ought to be more sympathetic to white suffering, or that white suffering is the epitome of human suffering. It is not just that the film reflects this attitude, but rather, that it performs and perpetuates this attitude as a work of art. *Blade Runner* suggests that we ought to care more about the white individuals in the film, and the hope that comes out of white suffering can be redemptive and beautiful. The non-white populations that populate the streets of *Blade Runner*'s Los Angeles barely share screen time.

Necro-temporality is also the time spent in toxic, violent, and surveilled spaces. While replicants move in and out of different areas, most of the spaces they move through or avoid death in are exposed to acidic rain and carcinogenic pollution. The sky is frequently dark from fumes and industrial waste, and the streets, alleys, hotels, and abandoned apartments show signs of advanced decay, suggesting poor structural integrity. They spend too much time in darkened alleys and abandoned apartments. These abandoned areas are also surveilled by corporations, which force the replicants to seek space between garbage cans or under garbage to hide. Like the remaining population on earth, replicants do not contribute to the booming economy of off-world colonies, and thus, the management and preservation of their life is not important.

Necro-temporality refers to the kind of time that emerges out of the logic of intensity behind neoliberalism. As I suggested above, *Blade Runner* incites its audience to live intensely through its message of hope. There is hope in the ability to remain resilient (e.g., to find love, redemption, and life even in horrific environmental conditions). The film's notion of hope is tied to the ability to take risks (Deckard risks the ire of the Tyrell Corporation and of other blade runners by not killing Rachel) and to accept the responsibilities that come with these risks. *Blade Runner* suggests that making these choices is what makes us human. The rhetoric of resilience in the film, then, reinforces the insecure conditions of neoliberal competition. These insecure conditions of neoliberal competition require individuals that want to live intensely because these individuals will not place blame on systems and governmentalities. Thus, these intense individuals thrive in insecure and potentially deadly conditions. Necro-temporality denotes the time of living intensely that is so integral to neoliberalism.

Spatial Pastiche and Blade Runner

For Jameson, pastiche is a defining trait of the postmodern condition. Pastiche is a consequence of "the disappearance of the individual subject." Pastiche is what happens when there are fewer examples of personal style. Pastiche mimics the past, but without humor. Jameson writes: "Pastiche is, like parody, the imitation of a peculiar mask, speech in a dead language: but it is a neutral practice of such mimicry...devoid of any laughter...Pastiche is thus blank parody..."[39] In the postmodern condition, cities become like museums in the sense that architecture becomes both preservative and derivative of the past. Architects use historicism to describe "the complacent eclecticism of postmodern architecture, which randomly and without principle but with gusto cannibalizes all the architectural styles of the past and combines them in overstimulating ensembles."[40] According to Jameson, historicism and nostalgia are both examples of pastiche.

Pastiche is also what Bruno, in her application of Jameson's postmodern aesthetics, calls an "aesthetic of quotations" that imitates dead styles.[41] Bruno uses pastiche as a model for analyzing *Blade Runner*'s architectural aesthetics. She finds in the film an aesthetic of decay that understands the city in terms of its disorder, disintegration, and pollution. The "idealized, aseptic technological order" of modernism with its skyscrapers, uber-modernity, and rationality is by and large absent in *Blade Runner*. Bruno describes bodily encounters with space as an artist or narrator setting or establishing the mood of a landscape or set. Toxic rain "completes the ambience" of postmodern decay, and its persistent falling "veil[s] the landscape of the city, further obscuring the neobaroque lighting."[42] Bruno's analysis examines the visual aesthetics of how *Blade Runner* represents a postmodern city and its postindustrial conditions. She looks for artistic markers, such as setting the mood, lighting, shadowing, and fashion. Bodies are less important than the spaces within which they exist. For example, she writes that it is "by garbage that Pris awaits J.S. Sebastian. A deserted neighborhood in decay is where Deckard goes…In an abandoned, deteriorating building, J.S. Sebastian lives…,"[43] thus suggesting that the bodies are mere decoration for the scenes of postmodern decay, disintegration, and waste. In Bruno's analysis, Pris and Zhora, along with women in the background, are "model[s] of the postindustrial fashion, the height of exhibition and recycling."[44] As representations of an aesthetic of recycling, another postmodern aesthetic, bodies are reduced to modeling fashion or wearable art.

Bruno's reading of *Blade Runner* perpetuates what Elizabeth Mahoney calls the "illusion of a depoliticized 'soft' postmodern landscape…" that "… in fact veils persistent and very real oppressive ideologies…beneath the façade of 'whimsy and pastiche' of the post-modern city, lie entrenched narratives of sexual difference."[45] While I would replace "entrenched narratives of sexual difference" with the conditions of competition, I think that Mahoney's critique of readings of postmodern aesthetics applies to Bruno's analysis.

The Cyberpunk City as a Necroscape

In this last section, I return to cognitive estrangement to think about the ways in which *Blade Runner*'s visual fashioning of a future Los Angeles furthers neoliberal resilience. *Blade Runner*'s visual fashioning of the cyberpunk city perpetuates notions of self-reliance and in some cases valorizes an intensified individualism. The city in *Blade Runner* is seemingly designed to support individual choice, freedom, and competition (all of which are considered positive by neoliberalism and many neoliberal subjects). But it has accelerated competition (visually we see the consequences of nurturing competition for competition's sake) to such a degree that individuals expect no support. They devalue and reject support, and if there are a few individuals who actually expect intervention (the replicants), they are eliminated from the population. *Blade Runner*'s Los Angeles is a necroscape because the power forces of this future city capitalize on the conditions of competition to both overtly and subtly kill unwanted bodies. The corporate state and its security apparatus, along with neoliberal governmental rationalities, marketize space. This means that all parts of the city are devoted to some form of competition, to the game of competing and self-entrepreneurship. In order to have free access to this space, populations must have human capital, and the will and resources to maximize this capital. Necro-temporality functions along with precarious urban design, state violence, and racism, all of which are part of the conditions of competition needed to ensure that some populations are more likely to live a quasi-life (fail to compete). Encounters between bodies, spaces, and power put some populations in a perpetual state of vulnerability, making replicants, and in many cases, the common masses of remaining humans, easier to exploit, kill, or let die. And, in a space where individuals expect no support and do not desire support, these individuals will not necessarily see these necroscapes as spaces of insecurity and death. Under conditions of competition for competition's sake, death will appear accidental, unavoidable, or as a matter of individual failure, but in reality, it is a result of living in toxic, insecure, and highly stressful spaces. It is no accident that the city is designed in such a way so as to keep illegal, racialized, and other unwanted bodies in insecure and toxic spaces. Living in insecure and toxic spaces makes the need for remaining resilient and for valuing resilience all the more necessary and all the more "natural." Thus, through making strange, through performing a cognitive estrangement with individual choice, freedom, and competition (notions that neoliberalism suggests have always been the way things are under neoliberalism), part of the goal here is to make it more difficult for neoliberalism to further its reality, and for neoliberalism to suggest that things have always been this way.

The corporate state in *Blade Runner* (and, I would suggest, neoliberalism in general) needs precarious urban design. One reason for this is that a failing urban environment combined with a deresponsibilized state and neoliberal values, such as self-reliance, resilience, and possessive individualism, help to make these notions necessary. As I suggested above, the individuals, such as the replicants, who resist total self-reliance, who resist devaluing and rejecting support, are eliminated from the population. *Blade Runner* shows this game of elimination, which is really a game of competition to see who will survive, through "the politics of verticality."[46] The higher a body is in the city, the more security and power it has over bodies on the ground, and the more likely it is to win in the game of survival. This verticality allows blade runners and other security forces to surveil the population from above, and clearly this gives blade runners an advantage in hunting illegal replicants. Living above ground also allows more competitive populations to avoid exposure to environmental toxins. For example, one scene in *Blade Runner* shows the computer inside a cop car purify its internal atmosphere as it increases its elevation. Blade runners are given special status in Los Angeles so that even a non-white blade runner like Gaff, a hybrid combination of different races, can avoid the streets. At one point, Deckard is shown looking over the city from the balcony of his high-rise apartment building, signaling his willingness to play the deadly game of competition, to do more to maximize his human capital, although this does not mean that if he fails to do his job (to compete successfully) he will be any more secure than replicants. Even blade runners must remain resilient and self-reliant if they wish to survive.

Individuals with low human capital are marked by earth status or street status, pointing to their economic, social, and genetic inferiority. They have no or very little human capital. J.F. Sebastian, a genetic designer for the Tyrell Corporation, informs Pris, a replicant, that he has Methuselah Syndrome, and later states, "My glands. They grow old too fast...I couldn't pass the medical exam."[47] Off world colonies can eliminate unwanted bodies through genetic testing. Many of the remaining humans are old, non-white, or hedonistic white bodies. Rather than focusing on making earth safer for the remaining humans, the corporate state has allowed large areas of Los Angeles to decay further. There is no need to engage in urban redesign because this insecurity makes neoliberal resilience necessary. Neoliberalism needs resilient and self-sufficient subjects and thus produces and relies on conditions that make resilience and self-sufficiency necessary.

Within the cyberpunk necroscape of Scott's *Blade Runner*, replicants dwell in insecure buildings, like J.S. Sebastian's abandoned apartment. Their insecure status makes replicants vulnerable to death by blade runners. Deckard enters Sebastian's abandoned apartment silently and without any barriers to his entrance. These spaces are not invaded since they cannot own property or lay claim to territory. Deckard then hunts for his prey in darkness and in silence. However, buildings like the Tyrell Corporation's mayanesque headquarters are managed by computer security systems that monitor and exclude bodies. Employees and guests can enter elevators, but they cannot go above a certain level in the building without permission from Eldon Tyrell himself. Roy only manages to get to the top floor by manipulating a current employee and outplaying Tyrell in chess. Leon, another replicant, enters the Tyrell Corporation, but in the waste disposal department, which is located in the bowels of the building. *Blade Runner*'s spatial presentation of seemingly über self-reliant individuals (the replicants) fighting to enter highly securitized buildings perpetuates narratives about corporate power and class struggles that valorize individualism. In the case of *Blade Runner*, however, this does not suggest that we ought to critique the system that makes these necrotic conditions necessary. *Blade Runner* does not intimate that individuals should advocate for a better state or a better system. Instead, it suggests that we should be a better individual. It is on us, as individuals, to live better, to want to live better in horrific conditions, or to be more resilient.

As a necroscape, the cyberpunk city makes familiar the ways in which state power kills and lets bodies die through city structures. Mike Davis, in his critique of *Blade Runner*, writes that it is derivative of Fritz Lang's 1927 film *Metropolis*. For Davis, Scott's film "captures ethno-centric anxieties about poly-glottism run amok,"[48] but it misses the real problems Los Angeles faces because it fails to capture the "spatial tendencies" of this modern city. The real spatial tendencies of Los Angeles are to securitize sectors, build upwards, and surveil, all of which serve the purpose of making some populations live. According to Davis, there are more technologically sophisticated high-rise buildings, more securitized, walled off neighborhoods, and more surveilled areas for the purpose of protecting some populations and making others vulnerable. However, his characterization of *Blade Runner* also misses how its corporate state, rather than invest in fixing current social conditions or decaying infrastructure, has turned to building upwards, securitizing buildings with advanced security systems, and surveilling large portions of the city, thus leaving racialized and illegal populations to dwell in the streets. Similar to the spatial apartheid that Davis aptly describes in his analysis of the real Los Angeles, individuals with more human capital in *Blade Runner* secure themselves, while allowing unwanted bodies to die in less secure spaces or be killed by security forces and while advocating for these conditions as a way to foster competition.

Conclusion

This chapter has shifted the focus away from the postmodern aesthetics of the city in *Blade Runner* and, instead, it has problematized the ways in which the temporalities and spatialities of neoliberalism have been defined (time-less time, schizophrenic temporality, and non-places) by showing that these notions capture only part of the critical possibility of the film. I have shown that Jameson's and Bruno's respective analyses of *Blade Runner* ignore the productive possibilities of a film like *Blade Runner*. I have advocated for a politically sensitive reading, not only of *Blade Runner*, but of cyberpunk as a genre. In addition, I have offered necro-temporality, necroscapes, and cognitive estrangement as tools for examining the ways neoliberal governmentalities make bodies vulnerable in cities through the cultivation of the spatio-temporalities of competition and the logic of intensity and self-reliance. This chapter has offered a way to think biopolitically about the cyberpunk film, *Blade Runner*. It is a way of thinking that makes the boundaries between science fiction and our neoliberal present less distinct, less concrete, and less taken-for-granted by highlighting the co-productive relationship between science fiction films and neoliberalism. Thinking biopolitically as a mode of cognitive estrangement means that we are able to consider the ways in which films produce and perpetuate the necrotic conditions of neoliberalism where living is defined as competing and dying is defined as a failure to compete or to remain resilient. This thinking biopolitically is the cognitive estrangement I advocated for above, since cognitive estrangement is a performance or process that seeks to make strange/fictive the social and economic status-quo. Cognitive estrangement, then, works towards making it more difficult for neoliberalism to function through its construction of reality as just the way things are and have always been. That is, cognitive estrangement, in this case, highlights the fact that perhaps resilience is not a practice or ontological state that we should so readily value. And, cognitive estrangement can point to the need to reevaluate our romanticization of an intensified individualism, since neoliberalism as a systemic reality perpetuates itself through the valuation of notions like resilience, individualism, intensity, and competition. Valuing resilience and wanting to be resilient justifies the harsh and degraded conditions that neoliberalism produces and depends upon. In the next chapter, I further the threads of resilience and intensity as I consider the possibilities of seeing beyond capitalism and I problematize the utopic vision of accelerationism. I highlight the necrotic conditions that underpin accelerationism, and I return to the figure of the cyborg as I examine the discourses of biohacking and the notion of the biohacker as an accelerationist/neoliberal subject.

Notes

1. Steven Shaviro, "Post-Cinematic Affect: On Grace Jones, Boarding Gate and Southland Tales," *Film-Philosophy*. 14, no. 1 (2010): 6.
2. Darko Suvin. *Metamorphoses of Science Fiction* (New Haven: Yale University Press, 1979).
3. Shaviro, "Connected," x.
4. Fredric Jameson, "The Cultural Logic of Late Capitalism," in *Postmodernism, or, The Cultural Logic of Late Capitalism* (Durham: Duke University Press, 1991), 44.
5. Ibid., 44, 54.
6. Shaviro, "Post-Cinematic Affect," 7.
7. Edward Said, *Orientalism* (New York: Vintage Books, 1979).

8. Donna Haraway, "A Cyborg Manifesto: Science, Technology, and Socialist-Feminism in the Late-Twentieth Century," *Simians, Cyborgs and Women: The Reinvention of Nature* (New York: Routledge, 1990), 149.
9. Steven Shaviro, "Post-Cinematic Affect," 42.
10. Marc Augé, *Non-places: Introduction to an Anthropology of Supermodernity*, trans. John Howe (London: Verso, 1995), 78.
11. *Boarding Gate* is a 2007 French film, directed by Olivier Assayas. The film follows a femme-fatale as she tracks a couple in Hong Kong. It is set largely within airports and high-end hotels and apartments.
12. Shaviro, "Connected," 130.
13. Jonathan Crary, *24/7: Late Capitalism and the Ends of Sleep* (London: Verso, 2013).
14. Manuel Castells, *The Rise of the Network Society, Vol. 1 of The Information Age: Economy, Society, and Culture* (Malden: Blackwell, 2000), 476.
15. Michel Foucault, *The History of Sexuality: Volume 1* (New York: Vintages Books, 1990).
16. Steven Shaviro, "The 'Bitter Necessity' of Debt," 7.
17. Giuliana Bruno, "Ramble City: Postmodernism and 'Blade Runner,'" *October* 41, (1987): 62.
18. Fredric Jameson, "The Cultural Logic of Late Capitalism," 63.
19. Ibid., 63.
20. Ibid., 4.
21. Achille Mbembe, "Necropolitics," *Public Culture* 15, no. 1 (2003): 40.
22. Giorgio Agamben, *Homo Sacer: Sovereign Power and Bare Life.* translated by Daniel Heller-Roazen (Stanford: Stanford University Press, 1998).
23. Carl Schmitt, *Dictatorship* (Malden, MA: Polity Press, 2014).
24. Mbembe, "Necropolitics," 17.
25. Jacques Lacan. *Écrits: A Selection*, trans. by Bruce Fink (New York: W. W. Norton and Company, 2004).
26. Bruno, "Rambling City," 70.
27. Ibid., 69.
28. Jameson, "Postmodernism," 72.
29. Ibid., 72.
30. Bruno, "Rambling City," 67.
31. Ridley Scott, *Blade Runner*, performed by Rutger Hauer (1982; Burbank; Warner Brothers), Streaming Video.
32. Bruno, "Rambling City," 70.
33. Ibid., 71.
34. Ibid., 70.
35. Scott, *Blade Runner*.
36. Mbembe, "Necropolitics," 21.
37. Scott, *Blade Runner*.
38. Ibid.
39. Jameson, "Postmodernism," 17.
40. Ibid., 65.
41. Ibid., 66.
42. Bruno, "Rambling City," 62.
43. Ibid., 64.
44. Ibid., 63.
45. Elizabeth Mahoney, "'The People in Parentheses': Space under pressure in the post-modern city," in *The Cinematic City*, ed. David B. Clarke (New York: Routledge, 1997), 171.

46. Eyal Weizman, "Introduction to The Politics of Verticality," *Open Democracy: Free Thinking for the World*, April 23, 2002, https://www.opendemocracy.net/ecology-politicsverticality/article_801.jsp.
47. Scott, *Blade Runner*.
48. Mike Davis, "Beyond Blade Runner: Urban Control, the Ecology of Fear," *Mediamatic Magazine*, 2009, http://www.mediamatic.net/6147/en/beyond-blade-runner-urban-control-1.

Bibliography

Agamben, Giorgio. "Threshold," in *Homo Sacer: Sovereign Power and Bare Life*. translated by Daniel Heller-Roazen, Stanford: Stanford University Press, 1998.

Augé, Marc. *Non-Places: Introduction to an Anthropology of Supermodernity*, translated by John Howe, London: Verso, 1995.

Bruno, Giuliana. "Ramble City: Postmodernism and 'Blade Runner,'" *October* 41, 1987: 62.

Davis, Mike. "Beyond Blade Runner: Urban Control, the Ecology of Fear," *Mediamatic Magazine*, 2009, http://www.mediamatic.net/6147/en/beyond-blade-runner-urban-control-1.

Foucault, Michel. "The Right of Death and Power Over Life," in *The Foucault Reader*, translated by Paul Rabinow, New York: Pantheon, 1984.

Haraway, Donna. "A Cyborg Manifesto: Science, Technology, and Socialist-Feminism in the Late-Twentieth Century," in *Simians, Cyborgs and Women: The Reinvention of Nature*, New York: Routledge, 1991: 127–148.

Jameson, Fredric. "The Cultural Logic of Late Capitalism," in *Postmodernism, or, The Cultural Logic of Late Capitalism*, Durham: Duke University Press, 1991: 1–54.

Mahoney, Elizabeth. "'The People in Parentheses': Space Under Pressure in the Post-Modern City," in *The Cinematic City*, edited by David B. Clarke, 171, New York: Routledge, 1997.

Mbembe, Achille. "Necropolitics," *Public Culture* 15, no. 1 (2003): 40.

Lacan, Jacques. *Écrit: A Selection*, translated by Bruce Fink, New York: W. W. Norton and Company, 2004.

Scott, Ridley. *Blade Runner*, performed by Rutger Hauer (1982; Burbank; Warner Brothers), Streaming Video.

Shapiro, Michael. *The Time of the City: Politics, Philosophy and Genre*, New York: Routledge, 2010.

Shaviro, Steven. Preface to *Connected: Or What It Means to Live in the Network Society*, Minneapolis: University of Minnesota Press, 2003.

—., "The 'Bitter Necessity' of Debt: Neoliberal Finance and the Society of Control," *Concentric: Literacy & Cultural Studies* 37, no. 1 (2010): 7.

—., "Post-Cinematic Affect: On Grace Jones, Boarding Gate and Southland Tales," *Film-Philosophy* 14, no. 1 (2010): 6.

Schmitt, Carl. *Dictatorship*, Malden, MA: Polity Press, 2014.

Suvin, Darko. *Metamophoses of Science Fiction*, New Haven: Yale University Press, 1979.

5 Reframing the Biohacker Within the Logic of Intensity

Introduction

In their accelerationist manifesto "#Accelerate: Manifesto for an Accelerationist Politics," Alex Williams and Nick Srnicek argue that we need to intensify the kind of human subject inherited from the Enlightenment: the human subject that masters itself and that masters nature. I use "intensify" here because I think Williams and Srnicek display an unreflexive neoliberal attitude towards self-mastery (a kind of uber-master), and they both align this self-mastery with a hacker ethic that is comfortable with complexity, artistry, and cunning, which is itself, an intensification of the "Promethean politics of maximal mastery."[1] Williams and Srnicek offer their accelerationist politics as an alternative to what they see as a failure in leftist politics to bring about the end of capitalism.[2] This is, of course, not new or unusual for accelerationism, and neither are, I would suggest, the problems with the "alternative" politics Williams and Srnicek offer. That is, accelerationism is in its numerous forms at best an unsatisfactory alternative to leftist politics and at worst a perpetuation of neoliberal resilience, as well as a push for a form of competition for competition's sake. I will return to this point later.

In this chapter, I suggest that the biohacker is the accelerationist subject Williams and Srnicek advocate for, and I argue that this accelerationist subject is, in the end, a neoliberal subject, like the self-monitoring cyborg, that fits easily within the conditions of competition. This chapter is, in part, an extension of Chapter 2. Like Chapter 2, I examine a concrete example of the cyberpunk subjectivities that I introduced in Chapter 1, but I focus here on the biohacker. Similar to the self-monitoring cyborg, the biohacker is subjectivized to live intensely and accept risk. The biohacker in its numerous forms reflects an underlying pure neoliberalism at work within accelerationism and its neoliberal governmentalities. I want to suggest here that, far from being an alternative to leftist politics, accelerationism may further the goals of neoliberalism in its desire to accelerate to a purified market space. The more recent forms of accelerationism, whether it be Nick Land's cyberpunk accelerationism or Williams and Srnicek's Promethean Mastery, are driven, I argue, by neoliberal notions of the individual, competition, and technology that, in the end, while perhaps not intended, further what I think is the ultimate goal of neoliberalism at its purest: to compete without limits, and in a manner whereby the only rules are those that protect competition. I want to push this point further to suggest that neoliberal accelerationism is another way of expressing a need for becoming more resilient in order to face the insecurities, the precarities, and the deprivations of neoliberalism. Neoliberalism is in many ways accelerative both in its abstractions and in its everyday material conditions. I am skeptical of the utopian possibilities of a "pure market accelerated out of capitalism altogether."[3] In the context of neoliberalism, this pure market beyond capitalism, which is a highly decoded and deterritorialized space, would not eliminate the competition for competition's sake that I am suggesting is part of neoliberal accelerationism.

The notions of self-mastery and resilience that permeate biohacker discourses suggest that as long as neoliberal governmentalities incite competition, whether through creative destruction, the compression of the present, urban design, 24/7 labor time, etc., subjects like the biohacker will live intensely in order to compete. In other words, the accelerationist politics that Williams and Srnicek offer remains a neoliberal politics. As a neoliberal subject, the biohacker hides the ways in which neoliberalism can govern without governing. That is, accelerationism's notions of self-mastery and resilience, similar to the notions of self-cultivation I examined in Chapter 2, reflect the kind of subjectivities (e.g., the possessive individual, the responsibilized-self, the risk-assessing individual, the resilient individual, the individual that lives intensely) that govern themselves according to the logics of competition and intensity. These subjects want to compete; they want to accept risk; they want to maximize their human capital. Its desire to maximize its human capital (e.g., to master and enhance the self, to overcome uncertainty and calculate risk, to conquer death, to endure the challenges of living in neoliberal capitalism, and to be exceptional/better than "normal" individuals), the fact that it has human capital, makes the biohacker/accelerationist imminently governable within neoliberalism.

Why not stop at an analysis of the self-monitoring cyborg, as offered in Chapter 2? What is different about the biohacker? To answer these questions, I want to emphasize a theme that runs through this study: living intensely/the logic of intensity. The biohacker is an intensification of the self-monitoring cyborg. The self-monitoring cyborg and the biohacker are not synonymous, but they do share similar practices, technologies, and discourses. They both join data technologies like the fitness tracker to their bodies to enhance their human capital. They both use fitness apps and communities to regulate their bodies and other bodies to maximize their human capital. They are both neoliberal subjects, although, as I will suggest in this chapter, in different ways. They both instrumentalize "self-cultivation." In the case of the biohacker, I would suggest it is "self-mastery" that is being cultivated in order to increase human capital. I think the distinction is important here because mastery again points to an intensification of cultivation. Mastery suggests a level of control, perhaps even a domination of the self, that cultivation does not. Biohackers often go beyond joining self-monitoring technologies to the body in order to master or enhance the self and the body, whether it is through permanently connecting a silicon transponder chip to a nerve, attaching magnets to the fingertips, taking nootropics, or injecting their own bone marrow into the body to create a youthful vigor.[4]

Thus, problematizing biohackers can highlight an intensification of biopolitical governmentalities and subjectivities under neoliberalism. This is not to suggest that these biopolitical governmentalities and subjectivities are not neoliberal governmentalities and subjectivities: they are both neoliberal and biopolitical, or perhaps neoliberal governmentalities are biopolitical governmentalities. Biohackers as neoliberal/biopolitical subjects compete to live. The life proper to neoliberalism, the life that is imminently governable/intelligible, is the life that competes to live (as we saw in previous chapters). Many biohackers embody the self-entrepreneur, and as such, subject aleatory events to calculability. That is, they turn the unknowability of the future into risk, and risk can be calculated in ways that unknowability cannot.[5] Biohackers often do the work of making live. They are often focused on maximizing their health, strength, biological age, and other forms of physiological performance, ultimately in order to compete. But this intensified focus on themselves also requires an individualized rationality that is always aware of the effects of living in particular environments and conditions. In some ways, they embody a neoliberal biopolitics.

This chapter will consider two different types of biohackers: DIYbiohackers and bulletproof biohackers. DIYbio-hackers advocate for and practice do-it-yourself genetic modification of humans and nonhumans, and some like Aaron Traywick advocate for and produce vaccinations, anti-biotics, and other medical treatments independent of large pharmaceutical companies.[6] By contrast, bulletproof biohackers often call for and practice taking control over human functionality/biology, often with the goal of slowing the ageing process and maximizing overall performance (e.g., using nutrition to enhance human biology, taking nootropics to enhance neural functioning, utilizing gene therapy, etc.). This is not to say that there are no overlaps among these two categories. These biohackers are all neoliberal subjects. As neoliberal subjects, DIYbio-hackers and bulletproof biohackers practice a subjectivity that is primarily focused on cultivating the self. Again, they are interested in maximizing human capital, although this is especially the case for the biohacker. "Human enhancement" is just another way of describing human capital.

I define biohackers as possessive individuals who embrace the responsibilized-self and the assumption of risk, and thus are often willing to take high risks (e.g., inject an untested herpes virus in front of a live audience, perform a fecal transplant, and perform other medical procedures without anesthesia) to maximize their human capital. And, as neoliberal subjects, biohackers are interested in becoming more resilient through human enhancement. This theme of resilience highlights the biohacker's, in its different forms, response to the conditions of competition within which biohackers compete to live and struggle to compete. By "compete to live," I mean that some biohackers, especially bulletproof biohackers, are already "living" since they often have more human capital (they are wealthy and have access to more resources) than DIYbio-hackers, and thus, they can compete more easily. I have argued in previous chapters that competing is equated with living under neoliberalism.

As this chapter unfolds, I show that the biohacker is an acceleration of the self-monitoring cyborg, which is perhaps yet another form of intensification. Earlier in this introduction, I pointed to some of the potential problems with accelerationism. The first section of this chapter explores some of the dominant accelerationists and their critiques. It suggests that accelerationism is, in the end, not a satisfactory answer to the left's failure to bring an end to capitalism, and that accelerationism offers some potentially problematic alternatives too (e.g., it is a form of neoliberal resilience). This section argues that the more recent accelerationist thinkers are, in the end, neoliberal theorists advocating for neoliberal conditions (e.g., competition and living life intensely) and for becoming more resilient. The second section analyzes biohacker discourses, including Dave Asprey's *BulletProof* website and its videos, and popular science and technology online magazine articles (*The Verge* and *H+ Magazine*, among others) about the biohacker, to argue that biohackers are driven by neoliberal accelerationism and that they are indeed neoliberal subjects. These individuals live intensely as they compete to live. Even though the biohacker's goal is to extend the life of the self, essentially to avoid death so that they can continue to compete, ironically, the biohacker's life is often like Roy Batty's short, yet intense life in *Blade Runner*. It is a life/light "that burns half as long, burns twice as bright."[7] And like Eldon Tyrell, the creator of replicants, there are neoliberal discourses (health, self-mastery/cultivation, the logic of intensity) that regularly remind biohackers that they are…"quite a prize…" and that they should "…[r]evel in your time."[8] As neoliberal subjects, biohackers maximize their resilience (their ability to endure the depravations of neoliberalism) the more intensely they live, since living intensely, whether it is self-practicing risky medical procedures, utilizing ever changing and unregulated body enhancements, or always adapting to the continually increasing speed of finance capitalism, allows individuals to make the most of their ability to compete. Yet, as I have argued throughout this study, we are all made more insecure under neoliberalism.

An Analysis of Neoliberal Accelerationism

Accelerationism is, at a basic level, a position that holds the view that, in order to move out of capitalism, we must go through it.[9] As Steven Shaviro puts it, for accelerationism to get out of capitalism, "we need to drain it to the dregs, push it to its most extreme point, follow it into its furthest and strangest consequences...the hope driving accelerationism is that, in fully expressing the potentialities of capitalism, we will be able to exhaust it and thereby open up access to something beyond it."[10] However, I suggest that this notion of accelerationism is not exactly what we are working with when it comes to neoliberal accelerationism. I define neoliberal accelerationism as the view that we must speed up capitalism in all of its capacities in order to get to a state of competition for competition's sake. The economization of the social, which is essentially an extension of economic thinking and practices into all parts of life, ensures that, even if this form of accelerationism were to exhaust capitalism and move past it, individuals would still live their lives as neoliberal subjects, as subjects that compete to live, as subjects that understand themselves in economic terms. Neoliberal accelerationism shows that it is not enough to speed up the processes of capitalism, since competition for competition's sake does not need capitalism to function. Neoliberalism's notion of competition currently seems to align well with finance capitalism, but it does not have to.

This section weaves in and out of two more recent forms of accelerationism (cyber accelerationism and promethean mastery) and their critiques, including Shaviro's and Noys's respective works on accelerationism, to highlight what I suggested above are the shortcomings of accelerationism, namely that it is at best an unsatisfactory answer to the Left's failure to end capitalism, and at worst, a form of "Deleuzian Thatcherism."[11] I then problematize Shaviro's accelerationist aesthetics by arguing that Shaviro is ultimately advocating for a form of neoliberal resilience. This section argues that accelerationism as a philosophy, political project, and even aesthetic perspective is a form of neoliberal resilience discourse, and that what appears to be other forms of accelerationism either serve the end goal of neoliberal accelerationism or are forms of neoliberal accelerationism.

Cyber-accelerationism

I want to begin with Shaviro's examination of Nick Land's version of accelerationism. Shaviro is correct to remain skeptical of the post-capitalist potentialities of Land's cyber accelerationism. He writes that Land's cyber accelerationism could just as easily end up as "a horrific intensification of 'actually existing' capitalist relations."[12] I would argue that this horrific intensification is, in fact, applicable to capitalist relations as they function within neoliberalism. Intensification informs neoliberal governmentalities. Intensification is a neoliberal governmentality, and it is an everyday form of neoliberal subjectivization (neoliberal subjects must live intensely and further must see living intensely as a virtue). Written in the 1990s, Land's accelerationism calls for the dissolution of the human through an integration with the technologies that have advanced within an information/finance capitalism. Land writes in "Machinic Desire" that the "Machinic revolution must therefore go in the opposite direction to socialistic regulation; pressing towards ever more uninhibited marketization of the processes that are tearing down the social field, 'still further' with 'the movement of the market, of decoding and deterritorialization' and 'one can never go far enough in the direction of deterritorialization: you haven't seen anything yet.'"[13] Land intensifies Gilles Deleuze and Félix Guattari's form of accelerationism.[14] According to Shaviro, Land's accelerationism argues that the violent and alien nature of capitalism, its destructive and inhuman forces, should be embraced as a vital force that, if allowed to reach its full potential, would dissolve the human ego and lead to an absolute deterritorialization,[15] which is for Land, a form of liberation.[16]

Land's cyber accelerationism, which was also part of the Cybernetic Culture Research Unit (CCRU),[17] claims that the full development of capitalism, its ability to be fully realized, was hampered by regulative policies that were often promoted by leftist politics. According to Land, the Left failed to end capitalism because its reactions to late capitalism did not allow its contradictions and capabilities to fully develop, and thus, to move beyond real subsumption. As Noys[18] makes clear in his book, *Malign Velocities*, a critique of accelerationism and its different manifestations, Land's accelerationist theory also aligned with what Foucault characterized as neoliberalism's permanent critique of the state. In *Malign Velocities*, Noys writes that Land's accelerationism, "despite its radicalized anti-humanism and inhuman immersive promise of capitalism exploding its own limits…,"[19] shares similarities with what are still current neoliberal claims, particularly "that capitalism wasn't really allowed to follow through."[20] Later in the text, Noys goes further to suggest that this accelerationism argues that "the acceleration of capitalism was held back by State spending and State regulation,"[21] and that "[i]t was a 'left' failure of nerve to go all the way to capitalism (and not all the way to the left…), that leaves us in the situation we find ourselves in."[22] For Land, this dissolution would alleviate some of the problems that have emerged and will continue to emerge out of replacing human labor with machine labor.

While I think that Land's cyber accelerationism emerged out of neoliberal conditions, I am hesitant to call it neoliberal accelerationism, at least in the way that I define it. As I suggested above, Land's criticism of the ways the state regulates capitalism aligns with neoliberalism's permanent critique of the state. Foucault writes in *"The Birth of Biopolitics"* that what develops under American neoliberalism is a kind of "economic positivism; a permanent criticism of governmental policy."[23] This is not quite the same kind of critique that emerges out of liberalism and its support of small government and minimum state interference. For liberalism, this continued criticism of big government stems from the belief that individuals are rational and should be left to themselves as autonomous, rational actors, particularly to make the kinds of rational decisions they may make in a market economy. Liberalism also assumes that competition is natural, and thus a state does not need to interfere with capitalism since the belief is that rational individuals will always regulate themselves.

Neoliberalism, however, does not assume that competition is natural. In fact, as Foucault makes clear, neoliberalism believes that competition is not natural at all. Thus, a state, even a powerful state, is justified for neoliberalism if it is needed to nurture and protect the necessary conditions for competition. Neoliberalism understands all state action in market terms. This means that, as Foucault puts it, "it involves anchoring and justifying a permanent political criticism of political and governmental action. It involves scrutinizing every action of the public authorities in terms of the game of supply and demand, in terms of efficiency with regard to the particular elements of this game, and in terms of the cost of intervention by the public authorities in the field of the market...the cynicism of a market criticism opposed to the action of public authorities."[24] Neoliberalism needs the state in order to protect competition. And, yet, at the same time, it is permanently engaged in a critique of the state. Neoliberal governmentalities, then, deresponsibilize the state's involvement with non-market concerns, and intensify individualism through responsibilization, risk-assessment, self-entrepreneurship, and so on.

Land's criticism of the state's unnecessary constraints on capitalism—the fact that, because of the state, capitalism was kept from reaching its own ends—aligns with this neoliberal critique of the state. For Land, the state is part of the "meat suit" that is holding individuals back from accelerating to a point beyond capitalism, to a point beyond the human even. This is what Noys calls Land's cyber accelerationism's "Deleuzian Thatcherism."[25] This means that, on the one hand, Land's cyber accelerationism is an expansion of Deleuze and Guattari's accelerative intensification of capitalism's deterritorialization and decoding. Land explicitly draws from Deleuze and Guattari's work in *Anti-Oedipus* when he writes that "you haven't seen anything yet" in terms of going still further towards "absolute deterritorialization."[26] On the other hand, his cyber accelerationism reproduces neoliberalism's permanent critique of the state,[27] and neoliberal governmentalities.

There are more problems with Land's accelerationism that need to be considered here. Land's accelerationism ignores the utopian fantasy of neoliberalism, and as hinted at above, it misses the ways in which neoliberalism can function without capitalism. The utopian fantasy of neoliberalism is, as I have already indicated, a deterritorialized competition for competition's sake, where the only rules are those that protect and perpetuate competition, where no one (human or not) is beholden to anyone else, and where the ultimate goal is to be "the last man standing."[28] I am not convinced that this fantasy is necessarily a capitalist one (Shaviro suggests that it is a purified state of capitalist accumulation,[29] and Noys believes that it is pure drive and accumulation[30]). Part of the goal of neoliberalism, its governmentalities, and its biopolitics is to subjectivize populations of individuals in such a way that individuals end up subjectivizing themselves as intensified, responsibilized, risk assessing, and possessive individuals who want to maximize their human capital in order to compete. These individuals do not necessarily want state interference except for where it may be needed to nurture and protect competition. The state, even a corporate state, could disappear and still these neoliberally subjectivized individuals would continue to compete, since it is not only what they ought to do, but it is what they "by nature" do.

Furthermore, competition is not necessarily instrumentalized to further capital accumulation. Individuals sell themselves because that is what individuals do. Selling the self and competing with others are now part of being in the world. Under real subsumption, they may serve capitalism. But, if we were to accelerate beyond capitalism, we would still be left with neoliberal ontologies/subjects that live by competing. Thus, the utopian goal of neoliberalism to compete for competition's sake could continue even in the absence of capitalism. Land writes in "Machinic Desire" that we must accelerate "towards ever more uninhibited marketization of the processes that are tearing down the social field."[31] As Noys makes clear, Land is advocating for a "purified capitalism,"[32] one that "would traverse to a pure market accelerated out of capitalism altogether."[33] I question whether a purified capitalism would really accelerate out of capitalism, since I think what we have seen, thus far, is a nightmarish intensification of capitalist relations. And, whether Land intends it or not, his cyber accelerationism could just as easily accelerate neoliberal accelerationism since this pure market could simply be a space and time where and when individuals compete for the sake of competing. Perhaps, Land's cyber accelerationism is, in the end, a kind of neoliberal accelerationism, or at the very least, Land's theory serves the accelerative ends of neoliberal accelerationism.

Promethean Mastery

In a different vein, Alex Williams and Nick Srnicek's "Manifesto for an Accelerationist Politics" is also a form of neoliberal accelerationism. Written in 2013, Williams and Srnicek argue that we need to distinguish between speed and acceleration. The distinction they draw here is not entirely convincing. They suggest that we can identify speed as that which describes what has happened and is happening under neoliberal capitalism, and that acceleration would describe a mastered control over the ways technology and human life would develop.[34] That is, under neoliberal capitalism, and they also seem to suggest under neoliberalism more generally, we "experience only the increasing speed of a local horizon, a simple brain-dead onrush…"[35] Capitalism, then, establishes the parameters within which we experience the speed up of time. The suggestion here is that we have no meaningful control over the way in which our lives develop in terms of time and space. For Williams and Srnicek, the solution is what they characterize as an acceleration of a "Promethean politics of maximal mastery over society and its environment."[36] This acceleration is not the same as speed for Williams and Srnicek, since to accelerate, here, implies a level of intentionality and agency that does not exist with the speed of capitalism. Williams and Srnicek also distinguish between the mastery they advocate and an Enlightenment mastery they characterize as outdated and inefficient in its ability to produce any "serious scientific understanding."[37]

Instead, they argue for a true, a real, an über self-mastery/mastery over nature. They claim that we need to produce a new, more complex mode of mastery that will use probability, a cunning rationality, geosocial artistry, and abductive experimentation[38] in order to produce the best ways of dealing with complexity. According to Williams and Srnicek, this self-mastery will produce the artistry, imagination, and vision necessary for a technological development that, if freed from capitalism, will help lead humankind to a post-capitalist society. But they are also careful to point out that they are not pushing for a techno-utopian vision, since they both argue that we need to accelerate technological development because it is "needed to win social conflict."[39] The assumption they make here is that, if we were to disentangle technology from capitalism, it would no longer be produced with its current built-in limitations, which includes built-in obsolescence. Thus, they want to complete what they call the "Enlightenment project of self-criticism and self-mastery," a project that they argue will lead us to "an alternative modernity that neoliberalism is inherently unable to generate."[40] I want to push back on the notion that neoliberalism is inherently incapable of generating this alternative modernity that Williams and Srnicek call for.

Williams and Srnicek conflate neoliberalism with late-capitalism, suggesting at times that they are either one and the same, or that they are both driven by the same goal (real subsumption). The neoliberalism and the kind of "alternative modernity" Williams and Srnicek are hoping for are not mutually exclusive. It may be the case that the kind of modernity they are advocating is antithetical to late-capitalism, but I would suggest that neoliberalism is responsible for the kind of intensified individualism that is, in part, driving Williams and Srnicek's accelerationism. The need to "win," to "be more decisive," "to quantify," to "be more cunning," to "be more creative," and to "maximize mastery" points to an underlying desire to compete and be competitive. It is as though Williams and Srnicek themselves are competing against other leftist political projects to see who will end capitalism first. Williams and Srnicek earnestly claim that "quantification is not an evil to be eliminated, but a tool to be used in the most effective manner possible. Economic modeling is—simply put—a necessity for making intelligible a complex world."[41] Their push to economize the social and capitalize on this economization actually does the work of neoliberalism even as they see themselves working against it. While they may not intend it, Williams and Srnicek are, in the end, promoting neoliberal accelerationism. The subject that is doing their version of accelerationism is a neoliberal subject. Theirs' is a neoliberal attempt to disrupt neoliberalism and capitalism, but one that in the end may further the end game of neoliberal accelerationism as a space and time of pure competition.

Accelerationist Aesthetics as a Form of Resilience

Shaviro does not see this "horrific intensification" as a reason to let go of accelerationism as a radical alternative to capitalism. Instead, he suggests that there is a middle ground to be found somewhere between Noys's negative critique of accelerationism and a full support of its efficacy. According to Shaviro, we should consider an accelerationist aesthetics before we can produce an effective accelerationist political project.[42] Still, Shaviro's aesthetic accelerationism perpetuates neoliberal discourses on resilience and the neoliberal governmentalities that incite, make valuable, and make necessary resilience and living intensely. A neoliberal subject that lives intensely is more resilient and is better at thriving, and in fact wants to thrive, in intensified conditions of competition.

I want to briefly return to Eldon Tyrell's advice to his prodigal son, Roy Batty, introduced above. Tyrell tells Roy that he is "quite a prize" and that he should "[r]evel in [his] time." I think this bit of "wisdom" precisely highlights neoliberal accelerationist positions on resilience and living intensely, and an examination of these positions can point to some underlying problems with the alternatives that accelerationism tries to offer. From Tyrell's perspective (after all, Tyrell is a type of neoliberal governmentality and power), Roy should feel lucky that he has the physical, emotional, and mental abilities to live intensely, always at the edge of burn out. In other words, according to Tyrell's neoliberal accelerationism, Roy ought to feel good about his natural abilities to remain resilient in the conditions of competition, and I am suggesting here that living intensely is a mode of resilience. That is, living intensely means that, as individuals engage in this intensification process, they maximize their abilities to survive/compete successfully in neoliberalism.

Resilience is, on one level, a virtue, seemingly independent of any particular economic, political, or social system: a virtue that, within neoliberalism and its accelerationism, Roy Batty ought to embrace.[43] On another level, resilience is a necessity for survival. Tyrell incites Roy to continue to live as intensely as possible, since this is really the only way Batty and his cohort can endure the depravations of neoliberalism. Living intensely as Roy does (he has "attacked ships on fire off the shoulder of Orion...watched C-beams glitter in the dark near the Tannhauser gate"[44]), and as many biohackers do, justifies the horrific conditions, in the case of Roy, and the insecure conditions, in the case of biohackers. That is, in the context of neoliberalism/accelerationism, the fact that Roy is a slave or that biohackers must always compete to live is treated as an experience that makes individuals more resilient, more competitive, just in general *more* of everything. It is not seen as a systemic problem. Their suffering is not suffering but is instead a sign of endurance and of gaining a competitive edge under neoliberalism.

Shaviro's aesthetic accelerationism thus echoes Tyrell's neoliberal resilience. In his book *No Speed Limits*, Shaviro argues that an aesthetic accelerationism may counter neoliberal capitalism's tendency to subsume any form of transgression or resistance by returning us to a post-capitalist aesthetic disinterest. He examines what he considers to be examples of accelerationist aesthetics, which for Shaviro primarily includes different works of science fiction. These accelerationist works themselves "provide a gloss on a situation that is already irreparable."[45] They "revel in the sleaze and exploitation that they so eagerly put on display. Thanks to their enlightened cynicism...they do not offer us the false hope that piling on the worst that neoliberal capitalism has to offer will somehow help to lead us beyond it."[46] Aesthetic accelerationism is, as Shaviro makes clear, capitalist realism,[47] since it fully embraces the fact that it assumes that there is no way out of capitalism, or that capitalism is our only option.[48] The point, it seems, for Shaviro, is that if we follow these "postures" we may find ourselves in a post-capitalist position of disinterest or detachment, although he is not quite clear on how this would work. Perhaps, at best what Shaviro offers us with his aesthetic accelerationism is a kind of "satisfaction and relief"[49] that we have "finally hit bottom, finally realized the worst."[50]

Later in his text, he argues that we need to accelerate an ethos of surplus and self-cultivation, since they are both an essential part of getting us to a post-capitalist society. But Shaviro is not clear on how we would disentangle self-cultivation from real subsumption under capitalism, nor is he clear on how we can move the aesthetic reflection of aesthetic accelerationism beyond "satisfaction [with] and relief"[51] at reaching rock bottom. Thus, Shaviro takes us on an accelerationist aesthetic trip that, in the end, suggests that we must learn to "live on in the face" of the depravations of neoliberalism and capitalism, with the hope perhaps of "finding a bounty" in times of scarcity.[52] He writes at one point that Paul Di Filippo's science fiction short story "Phylogenesis"[53] provides a way for us to adapt to capitalism and its monstrosities. A little later, he writes that, "[w]hen 'There is No Alternative'—when it no longer seems possible for us to defeat the monstrous invasion, or even to imagine things otherwise, this parasitic inversion is the best that we can do."[54] The parasitic inversion he is referring to here allows the humans of Di Filippo's world to remain resilient and endure the struggles of life. By feeding off the parasitic system that nearly destroys them, they learn how to survive.

Shaviro's position here is that humans must learn to be parasites on the body capital; we must learn how to mimic neoliberal capitalism. This is of course forgetting the fact that we already mimic capitalism as neoliberal subjects since very little escapes the economization of the social that Foucault speaks of in *"The Birth of Biopolitics"*. Shaviro concludes that, like the neohumans in Di Filippo's short story, we need to learn how to make the most of our lives. Despite the insecurities and deprivations that we face, Shaviro suggests that we need to accelerate an "ethos of abundance, generosity, and self-cultivation."[55] What alternative does Shaviro really offer us here? Shaviro's accelerationism is, in the end, another form of neoliberal resilience. In this final section of this chapter, I examine the biohacker as the parasite on the body capital that Shaviro encourages us to be. Biohackers mimic neoliberal capitalism (quantification, maximization, intensification, etc.). As neoliberal subjects, biohackers are making themselves better at surviving the horrors of neoliberalism and capitalism.

An Analysis of Biohacker Discourses

Over the past twenty years, the notion of the biohacker has become more and more popular. In 1998, Kevin Warwick, Deputy Vice-Chancellor of Research at Coventry University, began what some consider an early phase of the biohacker, his Project Cyborg.[56] Warwick, under anesthetics (not all biohackers have access to anesthetics), had a silicon chip responder implanted in the nerves of his left arm. That Warwick is sometimes called a biohacker, sometimes called a transhumanist, and at other times called a cyborg or even a post-humanist, points to perhaps what is, on some level, a lack of clarity and of a consistent definition/understanding about these terms and concepts. For the sake of my argument here, I accept the term biohacker since I think this word more clearly points to the ways in which DIY-biohackers and bulletproof biohackers are both neoliberal subjects. That is, various biohackers reflect the ways in which neoliberal subjects are accelerationists, and thus an analysis of biohacker discourses can further enlighten a critical reflection on neoliberal accelerationism. Biohackers are, in the end, making themselves more adept at competing. In this section, I analyze a number of biohacker discourses to illustrate how the biohacker is a neoliberal accelerationist and to show that living intensely as a form of resilience is tightly wound with neoliberal accelerationist subjectivity.

The Bulletproof Neoliberal as Biohacker

The first type of biohacker discourse that I analyze here is about the biohacker that has emerged along with neoliberal health and neoliberal resilience. Neoliberal health has individualized health by placing the responsibility on individuals, sometimes family units and/or communities, to prevent health problems and maximize bodily performances. At the same time, notions of self-mastery and self-cultivation still permeate tech companies that have produced quantitative data technologies, such as self-monitoring applications and wearable technologies. I suggest here that the biohacker has accelerated notions of self-cultivation to an intensified point of self-mastery. The word biohacker combines the act of hacking with human biology, which refers to figuring out creative, experimental, and new ways of fixing and mastering human biology. I focus in particular in this section on Dave Asprey because he embodies the neoliberal biohacker. He is the highly privileged, neoliberal subject, living on the edge of burn out, but who also has the resources to avoid burn out. Because his Bulletproof company offers numerous examples of neoliberal accelerationism, including an accelerationist/neoliberal resilience and a promethean self-mastery, his case is particularly instructive.

On the Bulletproof Blog, Asprey has a page that provides a brief statement about his life leading up to Bulletproof. The bio-statement indicates that Asprey has spent twenty years and one million dollars, thus far, to "hack his own biology."[57] That he has spent twenty years and one million dollars hacking his own biology lends his claims about biohacking, and even that biohacking is a science, some authority. But this time and money also highlight a level of persistence that is ideal for continually maximizing human capital, and, thus, for remaining competitive. We learn, for example, that Asprey lost 100 pounds "without counting calories or excessive exercise,"[58] something "normal" people have to do in order to lose weight, which is one among many ways to maximize human capital. Not only did Asprey more efficiently lose weight, but he also "used techniques to upgrade his brain and lift his IQ by twenty points...,"[59] and he "lowered his biological age while learning to sleep more efficiently in less time."[60] We also learn that, through these biohacks, Asprey has become "a better entrepreneur, better husband, and a better father."[61] In general, according to this bio-statement, it is clear that Asprey is better than most people, and he is better at being better. What might "seem impossible" for most people has become an everyday practice for Dave Asprey. He believes that biohacking is good, and he needs no motivation for maximizing his human capital. While notions of self-cultivation may haunt

Asprey's bio-statement and some of his videos, to call what he does, what he advocates for, or what he is self-cultivation would miss the ways in which pre-capitalist notions like self-cultivation are actually forms of mis-direction under real subsumption. Notions like self-cultivation take our attention away, turn our focus to other directions, or hide the ways in which Asprey's biohacking accelerates/intensifies self-cultivation to the point of self-mastery. This self-mastery is a kind of neoliberal and capitalist domination and production of the self that, for the most part, serves the purposes of neoliberal capitalism. Asprey's biohacking is an expression of one way in which the neoliberal subject subjectivizes itself.

Further down, Asprey's bio-statement tells us that his Bulletproof company can offer people different ways to take control of and improve their biochemistry that will help avoid "burning out, getting sick, and allowing stress to control your decisions."[62] Asprey may not overtly appeal to wealthier populations, but it is clear that the maximization practices he proposes for avoiding burn out require more resources than most populations of individuals can afford. He is speaking to a small, privileged audience. Asprey's "avoiding burnout" points to the logic of intensity that permeates many of the biohacker discourses, including those I examine throughout this section. In order to endure the depravations of neoliberalism and capitalism, individuals must become more resilient by, in this case, learning to live life intensely. If we think of life as an intense ride, then the nightmares of neoliberalism and capitalism become just another part of that intense ride, part of just the way things are, part of that which makes life exciting. Asprey's biohacker, then, can meet the challenges of the game of life that is competition by continually and obsessively performing at the highest levels, being better at everything, and upgrading their "performance in every aspect of life."[63] Being better at everything is the new normal of neoliberalism. Of course, Asprey, and others with his resources, do not have to actually "endure" the depravations of neoliberalism and capitalism because their resources allow them to avoid having to endure or struggle through life. As long as he has these resources, he can actually avoid burnout while still living the most intense and the best life.

In an "About Us" video on his Bulletproof site, Asprey provides the back story to the name Bulletproof. Asprey tells viewers that,

> Bulletproof isn't about being physically bulletproof. It is about being highly resilient. Who is bulletproof, Superman is bulletproof. That is what we all aspire to be, that is what we all have inside of us. We are humans of steal. We have a story in our head that doesn't give us credit. Everyone inside is like superman, but there are biological blocks to that. You literally are your own superhero. And that is why bulletproof is called bulletproof. Because we all want limitless strength and energy and we all want to be resilient so that we have the ability to handle whatever comes our way.[64]

Asprey claims that it is not about being physically bulletproof in the beginning of the video, but then when he describes what bulletproof means, being physically bulletproof does seem to be a part of the equation. He assumes that we all want to be like Superman in our desire to have limitless strength and energy. Here, he is regulating himself and he is participating in the regulation of others. It is clear that, for Asprey, there is a need for resilience. I believe that this need actually emerges out of the depravations of neoliberalism and capitalism. Asprey is answering this neoliberal call for resilience by inciting a desire for limitless strength and energy in order to become more resilient. Again, we must live more intensely (intensify strength, intensify energy, intensify mental performance, etc.) in order to become more resilient. Through this bulletproof biohacker discourse, there is a need to accelerate to a point of self-mastery that will enable subjects to live intensely, and thus to become more resilient, in the face of neoliberalism and its conditions of competition.

The DIY-biohacker and Its Accelerationism

As indicated above, another type of biohacker is the DIY-biohacker. This do-it-yourself biohacker is concerned with opening the ability to use creative experimentation on and manipulation of genes, both human and nonhuman, to the public. Anyone, for example, can buy a CRISPR kit[65] and use it at home to edit human and nonhuman genetic codes. In this section, I examine several articles that focus on a well-known DIY-biohacker named Josiah Zayner to, first, highlight the ways in which this form of biohacking is also neoliberal accelerationism, and, second, to make the case that this mode of biohacking is also about neoliberal resilience. Interspersed through this analysis of Zayner's DIY-biohacking are a few other articles and insights on DIY-biohackers.

Much of the accelerationist literature either diagnoses the present as a speeding up of time and compression of space, or it already accepts this speeding up of time and compression of space as conditions of neoliberal capitalism. Moreover, it argues that we need to push this acceleration even further to get to some kind of post-capitalist world. Of course, as I have suggested already, the post-capitalist world these literatures want to accelerate to will not necessarily be free of neoliberalism. The DIY-biohacker seems to sit somewhere in the latter camp of accelerationists. Alex Lash, in an article in *Exome* titled "No Self-Editing: Biohacker Josiah Zayner Can't Stop Living Out Loud,"[66] writes that "biohackers like Josiah Zayner think the world is moving too slowly."[67] This attitude is reminiscent of Williams and Srnicek's "Accelerationist Manifesto," since it is also coupled with the belief that more people should have access to technology or develop the technology in order to act upon their world, and further that this technology needs to be decoupled from capitalism. In other words, DIY-biohackers like Zayner are calling for a kind of "promethean politics of maximal mastery,"[68] as well as a kind of self-mastery and creative, playful mastery over nature.

Lash goes on to write that Zayner and others are "pushing the boundaries of do-it-yourself gene editing, for medicine, for exploration, and for fun, through home CRISPR kits and audacious displays of self-experimentation."[69] This characterization of the DIY-biohacker movement can turn our attention to a recent public moment when Zayner injected himself live at a conference in 2017 with DNA coding that was supposed to "enhance his muscles."[70] Soon after this live injection, Aaron Traywick, another DIY-biohacker, and recently deceased CEO of Ascendance Biomedical, injected himself on a livestream with an untested and unregulated herpes vaccine. Another biohacker for Ascendance Biomedical, Tristan Roberts, injected himself with an untested gene therapy treatment for HIV, and this was after he had stopped an FDA regulated treatment.[71]

In an article on his experience, Roberts expresses the belief that we are moving too slowly in our treatment of diseases. Roberts states that "we may be risk takers but we're not stupid..."[72] Later, after he started to experience possible side effects, he indicates that he is "fricking terrified it might be the plasmid..."[73] Yet, despite the risks, Roberts still wants to engage in these experiments. He wants to take risks. The willingness to take risks seems to be one way in which we can speed up the production of treatments. I want to suggest here that Roberts' desire to take risks is indicative of his neoliberal subjectivity. Zayner, Traywick, and Roberts are all examples of a neoliberal accelerationist subject. They are all driven by an obsessive desire to be the first one to produce a treatment, to compete with each other, and they are driven by a desire to live intensely through the assumption of risk and by taking risks (they assume an intensified risk), and they in many ways treat life as an intense ride. The belief that they ought to live intensely makes the risks they take in order to innovate seem independent of any particular system like neoliberalism. Their neoliberal subjectivity assumes that living is just naturally intense, dangerous, and precarious, and thus, that we ought to revel in our time, to speed-up innovation, creativity, experimentation to excess, though this accelerationism is not necessarily seen as excess since it is the norm of the logic of intensity.

In their "Accelerationist Manifesto," Williams and Srnicek work at describing the maximal mastery that is necessary to move to post-capitalism. Their description combines words like "improvisatory, creative, artistry" with words like "contingencies, executing, acting, cunning, and rationality" that are reminiscent of the ways in which biohackers describe their work, goals, rationales, and processes. As I have made clear, DIY-biohackers express a sense of urgency in their use of and production of technology and of the types of treatments, cures, and enhancements that may result from what they consider highly innovative, creative, exciting, and fun processes. Similar to Williams and Srnicek, DIY-biohackers want to decouple the capacities of technologies, specifically gene therapy technologies, from what they consider to be the restrictions of capitalism. In fact, DIY-biohacking as part of corporate endeavors often raises suspicions about whether what is being done is indeed DIY-biohacking.

In an interview published in February 2018 in *The Atlantic*, Zayner, who holds a PhD in biochemistry and biophysics, questions whether Traywick's livestreamed injection of the untested, unregulated herpes vaccination was an authentic form of DIY-biohacking. He characterizes what Traywick does as potentially sketchy, and he questions why Traywick sought out biohackers instead of "medical professionals, like a legitimate person would if you were trying to do some sort of legitimate research in gene therapies..."[74] Later, Zayner states that it "seems really strange and sketchy that someone [Traywick] would avoid legitimate researchers."[75] Zayner clearly sees a difference between what he does as a DIY-biohacker and what Traywick did as the CEO of Ascendance. Still, I am not so sure the distinction between the two is as clear as Zayner would like it to be.

Following this point, the interviewer Sarah Zhang suggests that they discuss the difference between biohackers and professional scientists. Zhang's distinction is that biohacking is "doing things that are hard or impossible to do inside the system...[t]hings like body augmentation or longevity research—things that the National Institutes of Health is uninterested in funding..." Zayner agrees with this characterization, and then they both suggest that what Traywick did was not DIY-biohacking because he was doing work that "lots of professional scientists are in fact working on under more regulation."[76] One of Zayner's more recent DIY-biohacking experiments was a fecal transplant, which is a procedure that is currently funded, performed by "legitimate" medical professionals, and highly regulated by the FDA.

What then is the difference between what Zayner did in performing his own fecal transplant outside of regulated laboratories (it was reported on in an article in *The Verge*)[77] and when Traywick livestreamed his injection of an untested herpes virus? Both are procedures that quite a few medical professionals are working on under regulation. Zayner suggests that part of the difference is that he is a social activist. When he livestreams his gene therapy injections or opens his experiments to journalists, he views these moments as ways to "make people think," as ways to normalize biohacking, as ways to encourage people to "play around" with gene therapy and other accelerative technologies. But, again, his language, his attitude, his persona, and his practices, in many ways, overlap with Traywick's justifications for calling what he does biohacking, and these overlaps suggest that Traywick's corporation was indeed doing DIY-biohacking.

Both wish to make cures and medical knowledge available to the public faster/earlier than they are usually made available (to make them immediately available). Both encourage creative play. Both want to push people to develop more and faster. Both are CEOs of corporations that encourage biohacking and that do biohacking. And both assume that this acceleration of knowledge production will motivate competition.[78] The DIY-biohacker's anti-capitalist ethos does not mean that DIY-biohackers are resistant to neoliberalism or that they offer an alternative political, ethical, social, or cultural project to neoliberalism. To the contrary, as I have been suggesting throughout this chapter, the DIY-biohacker is a neoliberal subject: a neoliberal accelerationist. Like Williams and Srnicek, they want to eliminate the "limitations imposed by capitalist society" for the purpose, I would argue, of accelerating competition. Once again, this is a competition that is decoupled from capitalism, and that, as such, pushes uninhibited innovation, which in turn pushes competition further.

Conclusion: Parasites on the Body Capital?

I want to end briefly by returning to Shaviro's aesthetic accelerationism. According to Shaviro, accelerationist art, which includes science fiction, and I would add cyber punk science fiction here, intensifies "the horrors of contemporary capitalism."[79] Through this intensification, accelerationist art provides us with a "kind of satisfaction and relief"[80] in its recognition that we can go no lower, that things cannot get any worse. Shaviro makes the case that this recognition of hitting rock bottom allows for a kind of detachment from the neoliberal present. For Shaviro, this detachment from the neoliberal present opens the door for laughter, for the ability to revel in hitting rock bottom, for a kind of realism that remains aware of itself. And all of this can lead us to an aesthetic that is outside of capitalism, or that is at least disinterested.[81] He goes on to write that any accelerationism worthy of being called accelerationism should contain an "ethos of surplus and self-cultivation."[82] I am not convinced that accelerationism, which draws from the systems (neoliberalism and capitalism) it wishes to accelerate out of, can draw upon or produce an ethos of surplus and self-cultivation, at least not the kind of ethos of surplus and self-cultivation Shaviro is pushing for. There may be an ethos of surplus here, but it functions as a way to remain resilient under neoliberalism. Instead of learning to live with less, let us make life a challenge, let us think of life as a game of competition, where we must take on the challenge of competing and of living intensely in order to remain competitive, and then we can learn to live with more and with being better, more/better strength, more/better intelligence, more/better life, more/better everything, so that we always remain ahead of the game of competition. This is the ethos of surplus under neoliberalism.

I remain skeptical that we can derive any viable political project from the logic of intensification that drives accelerationism. Aesthetic accelerationism does not really offer a detachment or distance from the neoliberal present. Or, if it does, it would be a detachment only for a privileged few. If anything, accelerationism's complicity, reveling, and lack of outrage in the depravations of neoliberalism reflect a fully immersed acceptance of living intensely. Of course, the neoliberal subjects of accelerationist works are complicit. Of course, they revel in their time. Of course, they lack outrage. They do not see the depravations of neoliberalism and often of capitalism as depravations at all, since a "normal" life is a life that is intense, dangerous, and precarious. These depravations of neoliberalism and neoliberal capitalism, whether it is the desolation of the environment, of work, or of the body, are just the way things are. And thus, living intensely enables a kind of resilience that makes these depravations endurable, and sometimes even exciting or pleasurable. Biohackers gain endless pleasure out of living intensely.

Notes

1. Alex Williams and Nick Srnicek, "#Accelerate: Manifesto for An Accelerationist Politics," in *#Accelerate#: The Accelerationist Reader* (Windsor, UK: Urbanomic, 2014), 361.
2. Ibid.
3. Benjamin Noys, *Malign Velocities: Accelerationism and Capitalism* (Winchester, UK: Zero Books, 2014), 55.
4. Ben Popper, "Cyborg America: Inside the Strange New World of Basement Body Hackers," *The Verge*, August 18, 2012, https://www.theverge.com/2012/8/8/3177438/cyborg-america-biohackers-grinders-body-hackers.
5. Steven Shaviro, *No Speed Limit: Three Essays on Accelerationism* (Minneapolis: University of Minnesota Press, 2014), 11.
6. Popper, "Cyborg America."
7. Ridley Scott, *Blade Runner*, performed by Joe Turkel (1982; Los Angeles: Warner Brothers, 2007), DVD.
8. Scott, *Blade Runner*.
9. Shaviro, *No Speed Limits*, 2.
10. Ibid., 3.
11. This is the idea that cyber accelerationism's critique of the state as holding individuals back from accelerating to a point beyond capitalism parallels the permanent neoliberal critique of the state. I elaborate on this point later.
12. Ibid., 20.
13. Nick Land, "Machinic Desire," in *Fanged Noumena: Collected Writings 1982–2007*, eds. Robin Mackay and Ray Brassier (Falmouth, UK: Urbanomic, 2011), 480.
14. Noys in *Malign Velocities* and Shaviro in *No Speed Limit* characterize Deleuze and Guattari's claim that "…perhaps the flows are not yet deterritorialized enough, not decoded enough, from the viewpoint of a theory and practice of highly schizophrenic character. Not to withdraw from the process, but to go further…the truth is that we haven't seen anything yet,"

as a form of accelerationism that shaped what accelerationism has become in the last 28 years. See Gilles Deleuze and Félix Guattari, *Anti-Oedipus*, trans. Robert Hurley, Mark Seem, and Helen R. Lane (Minneapolis, MN: University of Minnesota Press, 1983), 239–240.
15. Gilles Deleuze and Félix Guattari, *Anti-Oedipus*, trans. Robert Hurley, Mark Seem, and Helen R. Lane (Minneapolis, MN: University of Minnesota Press, 1983).
16. Ibid., 14.
17. The CCRU was a transdisciplinary student run collective. It was founded in 1995 and was associated with the University of Warwick's philosophy department.
18. Noys is credited with having coined the term accelerationism in his 2010 book *The Persistence of the Negative*; See Benjamin Noys, *The Persistence of the Negative: A Critique of Contemporary Continental Theory* (Edinburgh: Edinburg University Press, 2010).
19. Benjamin Noys, *Malign Velocities: Accelerationism and Capitalism* (Winchester,UK: Zero Books, 2014), 56.
20. Ibid., 56.
21. Ibid., 56.
22. Ibid., 56.
23. Michel Foucault, *The Birth of Biopolitics: Lectures at the Collège de France, 1978–1979* (New York: Palgrave MacMillan, 2004), 247.
24. Ibid., 246.
25. Noys, *Malign Velocities*, xi, 56.
26. Ibid., 2.
27. I accept Foucault's definition of the state as "nothing else but the mobile effect of a regime of multiple governmentalities." Michel Foucault, "*The Birth of Biopolitics*," 77.
28. Shaviro, *No Speed Limits*, 23.
29. Ibid., 23.
30. Noys, *Malign Velocities*, 52.
31. Land, "Machinic Desire," 480.
32. Noys, *Malign Velocities*, 55.
33. Ibid., 55.
34. Williams and Srnicek, "#Accelerate," 352.
35. Ibid., 352.
36. Ibid., 360.
37. Ibid., 361.
38. Ibid., 361.
39. Ibid., 356.
40. Ibid., 362.
41. Ibid., 356.
42. Shaviro, *No Speed Limits*, 20–21.
43. See Brad Evans and Julian Reid, *Resilient Life: The Art of Living Dangerously* (Cambridge, UK: Polity Press, 2014); and David Chandler and Julian Reid, *The Neoliberal Subject: Resilience, Adaptation and Vulnerability* (Lanham, MD: Rowman and Littlefield, 2016).
44. Scott, *Blade Runner*.
45. Shaviro, *No Speed Limits*, 37.
46. Ibid., 35–36.
47. Mark Fisher, *Capitalist Realism: Is There No Alternative?* (Hants: Zero Books, 2009).

48. Shaviro, *No Speed Limits*, 39.
49. Ibid., 35.
50. Ibid., 35.
51. Ibid., 35.
52. Ibid., 56.
53. Paul Di Filippo, "Phylogenesis," in *Babylon Sisters and Other Posthumans* (Canton, OH: Prime Books, 2002).
54. Shaviro, *No Speed Limits*, 59.
55. Ibid., 60.
56. Kevin Warwick, "Project Cyborg 1.0," *Kevin Warwick*, accessed June 29, 2020, http://www.kevinwarwick.com/project-cyborg-1-0/.
57. Dave Asprey, "Dave Asprey, Founder and CEO of Bulletproof," accessed June 29, 2020. https://blog.bulletproof.com/about-dave-asprey/.
58. Ibid.
59. Ibid.
60. Ibid.
61. Ibid.
62. Ibid.
63. Ibid.
64. Dave Asprey, "The Story Behind the Name: Bulletproof," accessed June 29, 2020. https://www.bulletproof.com/pages/about-us.
65. CRISPR stands for Clustered Regularly Interspaced Short Palindromic Repeats. It is a type of gene editing technology; see "Genetics Home Reference," *U.S. National Library of Medicine*, accessed January 15, 2019, https://ghr.nlm.nih.gov/primer/genomicresearch/genomeediting.
66. Exome is part of Xconomy, an online news source devoted to the "exponential economy."
67. Alex Lash, "No Self-Editing: Biohacker Josiah Zayner Can't Stop Living Out Loud," *Exome* (2018), accessed June 29, 2020, https://www.xconomy.com/san-francisco/2018/03/13/no-self-editing-biohacker-josiah-zayner-cant-stop-living-out-loud/.
68. Williams and Srniceck, "#Accelerate," 360.
69. Lash, "No Self-Editing," 2018.
70. Sarah Zhang, "A Biohacker Regrets Publicly Injecting Himself With CRISPR," *The Atlantic*, February 20, 2018, https://www.theatlantic.com/science/archive/2018/02/biohacking-stunts-crispr/553511/.
71. Jessica Lussenhop, "Why I Injected Myself with an Untested Gene Therapy," *BBC News*, November 21, 2017, https://www.bbc.com/news/world-us-canada-41990981.
72. Ibid.
73. Ibid.
74. Zhang, "A Biohacker Regrets," https://www.theatlantic.com/science/archive/2018/02/biohacking-stunts-crispr/553511/.
75. Ibid.
76. Ibid.
77. Arielle Duhaime-Ross, "A Bitter Pill," *The Verge*, February 4, 2016, https://www.theverge.com/2016/5/4/11581994/fmt-fecal-matter-transplant-josiah-zayner-microbiome-ibs-c-diff.
78. Ibid.
79. Shaviro, *No Speed Limits*, 35.
80. Ibid., 35.
81. Ibid., 40.
82. Ibid., 55.

Bibliography

Asprey, Dave. "The Story Behind the Name: Bulletproof," accessed September 17, 2018. https://www.bulletproof.com/pages/about-us.
—., "Dave Asprey, Founder and CEO of Bulletproof," accessed September 17, 2018. https://blog.bulletproof.com/about-dave-asprey/.
Chandler, David and Reid, Julian. *The Neoliberal Subject: Resilience, Adaptation and Vulnerability*, Lanham, MD: Rowman and Littlefield, 2016.
Deleuze, Gilles and Guattari, Félix. *Anti-Oedipus*, translated by Robert Hurley, Mark Seem, and Helen R. Lane. Minneapolis, MN: University of Minnesota Press, 1983.
Di Filippo, Paul. "Phylogenesis," in *Babylon Sisters and Other Posthumans*, Canton, OH: Prime Books, 2002.
Duhaime-Ross, Arielle. "A Bitter Pill," *The Verge*, February 4, 2016, https://www.theverge.com/2016/5/4/11581994/fmt-fecal-matter-transplant-josiah-zayner-microbiome-ibs-c-diff.
Evans, Brad and Reid, Julian. *Resilient Life: The Art of Living Dangerously*, Cambridge, UK: Polity Press, 2014.
Foucault, Michel. *The Birth of Biopolitics: Lectures at the Collège de France, 1978–1979*, New York: Palgrave MacMillan, 2004.
Land, Nick. "Machinic Desire," in *Fanged Noumena: Collected Writings 1982–2007*, edited by Robin Mackay and Ray Brassier. Falmouth, UK: Urbanomic, 2011.
Lash, Alex. "No Self-Editing: Biohacker Josiah Zayner Can't Stop Living Out Loud," *Exome* (2018): https://www.xconomy.com/san-francisco/2018/03/13/no-self-editing-biohacker-josiah-zayner-cant-stop-living-out-loud/.
Lussenhop, Jessica. "Why I Injected Myself with an Untested Gene Therapy," *BBC News*, November 21, 2017, https://www.bbc.com/news/world-us-canada-41990981.
Noys, Benjamin. *Malign Velocities: Accelerationism and Capitalism*. Winchester, UK: Zero Books, 2014.
Popper, Ben. "Cyborg America: Inside the Strange New World of Basement Body Hackers," *The Verge*, August 18, 2012. https://www.theverge.com/2012/8/8/3177438/cyborg-america-biohackers-grinders-body-hackers.
Scott, Ridley. *Blade Runner*, performed by Joe Turkel (1982; Los Angeles: Warner Brothers, 2007), DVD.
Shaviro, Steven. *No Speed Limit: Three Essays on Accelerationism*. Minneapolis: University of Minnesota Press, 2014.
Warwick, Kevin. "Project Cyborg 1.0," *Kevin Warwick*, accessed November 6, 2018, http://www.kevinwarwick.com/project-cyborg-1-0/.
Williams, Alex and Srnicek, Nick. "#Accelerate: Manifesto for an Accelerationist Politics," in *#Accelerate#: The Accelerationist Reader*. Windsor, UK: Urbanomic, 2014.
Zhang, Sarah. "A Biohacker Regrets Publicly Injecting Himself with CRISPR," *The Atlantic*, February 20, 2018. https://www.theatlantic.com/science/archive/2018/02/biohacking-stunts-crispr/553511/.

Conclusion
Defamiliarizing Neoliberalism Through Cyberpunk Science Fiction

Introduction

What does it mean to suggest that cyberpunk science fiction is accelerationist art? It may mean that it "intensifies the horrors of contemporary capitalism."[1] And that, by reading and thinking about cyberpunk, we derive satisfaction and relief[2] through its ability to help us see that things cannot get any worse than they already are. For Shaviro, this recognition of hitting rock bottom allows for a kind of detachment from the current neoliberal reality that may allow us to think outside of capitalism, or for an "ethos of surplus and self-cultivation," as he puts it, that would help induce an aesthetic that is beyond capitalism. As I suggested in Chapter 5, this study remains skeptical of the critical post-capitalist possibilities of accelerationism and of accelerationist art like cyberpunk. The ethos of surplus and self-cultivation that Shaviro considers is instead a way to remain resilient in the face of neoliberalism and its horrors. The ethos of surplus and self-cultivation abides by a logic of intensity, and this ethos is a form of neoliberal resilience. The logic of intensity as a form of neoliberal resilience, whether it is found in cyberpunk's amplification, in the self-monitoring cyborg's living on the edge of burn out, or in the biohacker's living intensely, has been traced throughout this study. In this final section, I return to some of the core considerations of this study. First, I look back at the ways in which I defamiliarized neoliberalism's reality construction. I then briefly consider how this study rethinks biopolitics in its neoliberal form. And, finally, in this conclusion, I return to thinking about the logic of intensity as a form of neoliberal resilience as it weaves through the narrative flow of this study. Here, I briefly revisit necro-temporality and necroscapes as the time and space(s) of the logic of intensity and neoliberal resilience.

Defamiliarizing Neoliberalism's Reality Construction

As I indicated in the Introduction, this study aims at making it more difficult for neoliberalism to produce and nurture a reality that "already exists." In other words, this study makes strange and thinks strangely about neoliberal values, governmental rationalities, and modes of knowledge production that are often assumed to be just the way things are. Making strange, thinking strangely, or cognitive estrangement are all forms of defamiliarization. This study has defamiliarized neoliberalism's reality construction through its simultaneous engagement with cyberpunk science fiction and neoliberalism. This is not to suggest that I have only engaged with neoliberal science or that neoliberalism is only fictive through the kinds of science it offers. Rather, I have juxtaposed cyberpunk science fiction with neoliberalism in order to generate epistemological interferences that make it harder to ignore neoliberalism's narrative and its contingent qualities. I use the term "juxtapose" here to mean, in part, that thinking about cyberpunk and neoliberalism in the same space, at the same time, and, at times, in the same way, can force readers to see both a work of literature and a systemic reality in a different light. Through my use of "juxtapose," I wish to imply that a certain level of violence is being done to cyberpunk and neoliberalism. I have violated the assumed differences or boundaries between a work of art in the form of cyberpunk and some of the systemic conditions of the present, namely neoliberalism and capitalism.

Thus, as I have argued throughout this study, rather than thinking about cyberpunk as a work of art that functions as a mirror image of the real (the real here is often the postmodern condition of late-capitalism), this study has looked at the co-productive relationship between cyberpunk and neoliberalism. That is, I have emphasized what cyberpunk might do, often in spite of itself, such as when it produces or engages in neoliberal governmental rationalities and subjectivities, when it helps to defamiliarize neoliberal values and the conditions of competition, and/or when it frames how or whether we comprehend hyperobjects of late-capitalism. Unlike much of the foundational literature on cyberpunk, I have worked here against the notion that cyberpunk serves as evidence of the fact that Jameson's, Harvey's, and Bruno's (among others) respective diagnoses about the spatial and temporal conditions of late capitalism are real. Again, I think that such an intervention is important because, rather than working towards diagnosing or reflecting on the present, this study has done more than describe the postmodern condition and lament the loss of a self rooted in instrumental rationality. Instead, if we think that cyberpunk is a productive force in neoliberalism, that it furthers neoliberal reality, or simply that cyberpunk is neoliberalism or does neoliberalism, in part, through amplification and cyborganization, then the historical, social, political, and economic contingencies of neoliberalism can become more critically apparent.

The suggestion here is that neoliberalism is also cyberpunk. In other words, by violating the assumed boundaries between literary genres, in this case cyberpunk, and neoliberal truths, we can reveal the narratives of neoliberalism that associate good character with individual responsibility and rational risk assessment, that naturalize the entrepreneurial self, that de-responsibilize state agents, and that, at the same time, highlight underlying necrotic conditions that foster these narratives. This means that the everyday self-monitoring practices of a cyberpunk/neoliberal subjectivity like the self-monitoring cyborg are not about self-cultivation, rational individualism, or choice. Nor are the everyday biohacking practices of a cyberpunk/neoliberal subjectivity like the biohacker about self-mastery, living intensely, or a do-it-yourself ethos. Instead, once again, this study has sought to make strange these seemingly benign discourses, rationalities, and subjectivities in order to highlight the role they play in neoliberalism's ability to perpetuate itself. Neoliberalism, like biopolitics, functions through the elimination of individuals and groups that pose a threat to the proliferation of healthy populations. But, for neoliberalism, healthy means competitive, resilient, and intense.

Neoliberal Biopolitics and Necro-temporality

The way I have theorized biopolitics in this study is in part a response to Elizabeth Povinelli's intervention in the literature on biopolitics. In an interview about her book *Geontologies: A Requiem to Late Liberalism*, Povinelli argues that the theorization of biopolitics has been too focused on spectacular death and biological racism, and that this zero-sum focus misses ordinary suffering and the common deaths that are the result of the kinds of enduring and suffering she conceptualizes. For Povinelli, then, biopolitics is insufficient to help us to understand the ways that power functions in late liberalism.[3] This study has offered a theorization of biopolitics that apprehends death in terms of ordinariness, and it has argued that biopolitics can still offer us insights into the ways in which neoliberalism functions and constructs its reality. As I have suggested through the previous chapters, the ways in which power functions within biopower is intensified in the context of neoliberalism. Biopolitics defines and controls life through an integration of economic calculation and economic rationalities, which I think still makes it a useful analytic for understanding how some neoliberal governmental rationalities function. After all, under neoliberal conditions, populations of individuals become intelligible through economic calculation.

I push back against Povinelli's notion that biopolitics is defined by "outright extinction" rather than "perseverance, endurance, effort, and precarious survival."[4] Perseverance, endurance, effort, and even precarious survival describe the kinds of resilience that are made necessary by and under neoliberalism. With resilience, subjects can compete and seek to live the most intense lives as they work without benefits and security, maximize their human capital, and embrace flexible work. The space and time of neoliberalism are necrotic, in part, because of a hostility towards the social and political ontologies that do not produce and nurture a logic of competition. Neoliberal governmental rationalities perpetuate and maintain competition and then de-responsibilize their role in producing these conditions. Necro-temporality accounts for the ways in which the time and space of neoliberalism is necrotic. Since neoliberalism intensifies responsibility, individuals are more and more on their own, while, at the same time, they have no other option but to embrace this neoliberal reality. Necro-temporality and necroscapes describe and also problematize the temporalities and spatialities that a generalized mode of existence defined by competition with others has made possible. In this study, I have understood necro-temporality as the condition of insecurity and, further, of being responsible for one's insecurity, as the naturalization of dying from endemic illnesses caused by living on the edge of burn out (living intensely), and crucially as a time when neoliberal subjects embrace notions like "living life to the fullest" or "seizing the day" as forms of resilience. Necro-temporality is a biopolitical technique for neoliberal governmentalities to manage and control how we define or treat death, and deal with individuals with less human capital. In the end, necro-temporality makes it easier for state agents, governmental rationalities, institutions under neoliberalism, in other words, for neoliberal hegemony, to dismiss, ignore, and rationalize the death of some populations as natural, inevitable, unsolvable, or as a failure of character. Necro-temporality is thus neoliberal time. It is the time that has emerged out of the logic of intensity and competition, both of which are key to neoliberalism. Under necro-temporality, individuals want to live intensely, and they seek to thrive in risky and deadly conditions. The notion that all time is work time is a product of necro-temporality. Necro-temporality is a time of living intensely that is foundational to neoliberalism. Neoliberal subjects like the self-monitoring cyborg and the biohacker, at varying degrees, are subject to necro-temporality as they live always on the edge of burnout. Likewise, there are plenty of individuals who must endure a life of living beyond burnout.

A Final Word on Necropolitics

The goal of my attempt at disentangling Michel Foucault's biopolitics from Achille Mbembe's necropolitics, as I have done in Chapter 3, has not been to diminish Mbembe's and other post-colonial contributions to contemporary studies of biopolitics. I do not doubt that Foucault's biopolitics is in some ways limited in what it can tell us about the temporalities, spatialities, and the nature of death and destruction in colonial and post-colonial spaces, as many have argued. But my aim has been different. It has been to create a conceptual bridge between biopolitics and necropolitics, and in particular to suggest that, since killing is a positive condition of biopolitics, biopolitics does not necessarily need an addendum or a corrective. In other words, as I have shown in much of this study, we may not need necropolitics as a way to describe what biopolitics can or cannot tell us about power and death-making since what I have suggested throughout this study is that biopower, by its very definition, kills in order to make live. We thus can conceptualize killing or death-making as existing along a spectrum (it is not always spectacular, and it is not always ordinary either). The kind of death-making that I have considered in this study is not direct though. Sometimes, it is indirect. Often, it does not appear as death-making: it is the product of a death-producing system, but it cannot always easily or directly be traced back to this particular system. It is through disentangling biopolitics from necropolitics that I have shown, first, that theorizing biopolitics need not be defined by "outright extinction," and, second, that death-making operates as a spectrum within Foucault's biopolitics framework.

Notes

1. Steven Shaviro, *No Speed Limit: Three Essays on Accelerationism* (Minneapolis: University of Minnesota Press, 2014), 35.
2. Ibid., 35.
3. Elizabeth Povinelli, "An Interview with Elizabeth Povinelli: Geontopower, Biopolitics and the Anthropocene," eds. Mathew Coleman and Kathryn Yusoff. *Theory, Culture & Society* 34, no. 2–3 (2017): 170.
4. Ibid., 170.

Bibliography

Povinelli, Elizabeth. "An Interview with Elizabeth Povinelli: Geontopower, Biopolitics and the Anthropocene," edited by Mathew Coleman and Kathryn Yusoff. *Theory, Culture & Society* 34, no. 2–3 (2017): 169–185.

Shaviro, Steven. *No Speed Limit: Three Essays on Accelerationism*. Minneapolis: University of Minnesota Press, 2014.

Index

accelerationism 17–19, 25, 37, 104, 107–114, 116–119, 121–122, 124–125, 129: accelerationist 104, 107–108, 110–112, 114, 116–119, 121–122, 124–125, 129; aesthetics 111, 116–117; neoliberal 25, 108, 110–112, 114, 116, 118–119, 121; as Promethean Mastery 107–108, 111, 114–115, 119, 122
Agamben, Giorgio 8, 66, 95
Altered Carbon 29
amplification 17, 19, 23–25, 27–28, 33, 84, 129–130
Asprey, Dave 110, 119
Augé, Marc 87–88

Basis 51–52
Baudrillard, Jean 11, 95
Biohacker 80, 104, 107–110, 117–125, 129, 131–132; DIY biohacker 109–110, 118, 121–124
biopolitical subject 109
biopolitics 4–10, 13, 16–19, 23, 25, 27, 29, 32, 45, 47–49, 65–67, 69, 71–73, 75–81, 89–90, 93–94, 96, 109, 113–114, 118, 129, 131–133; biopower 1, 3, 5–7, 11, 13, 18, 30, 45, 48–49, 51–52, 58, 60, 65–75, 77, 80–81, 89–90, 93, 131, 133; liberal 5; neoliberal 4–5, 10, 17, 23, 27, 32, 65, 72, 75, 78–81, 89, 96, 109, 131
Blade Runner 15, 18, 29–31, 33–34
Bruno, Giuliana 3, 11–12, 88, 90–92, 95–97, 100, 103, 130
Bukatman, Scott 15–16, 24, 34–35
bulletproof biohacker 109–110, 118–121

Cacho, Lisa Marie 29–31, 75–76, 79–80
capitalism 3, 6, 11–13, 15, 17–18, 20, 24, 31–33, 36, 41–42, 60, 80, 86, 88, 90–91, 92, 94, 104, 107–108, 110–118, 120–125, 129–130; advanced 12–13; corporate 23, 30; finance 110–112; industrial 6; late 3, 6, 11–13, 15, 41–42, 86, 88, 90–91, 112, 115, 130; logic of 3, 11, 15, 88, 90–91; multinational 13; neoliberal 17, 92, 115, 117–118, 120–121, 125; post 122; purified 114; transnational 33
city 3–4, 9–12, 15–16, 25–28, 30–33, 80, 82, 88, 90–91, 98–103
Clausewitz, Carl von 74
cognitive estrangement 86–87, 90, 92, 100–101, 103–104, 130
cognitive mapping 14, 86–87, 90
competition 1–3, 5–6, 9–11, 13–14, 19, 23, 25–26, 28–29, 31–32, 36, 41–43, 45–47, 52, 56, 65–66, 75, 77–78, 80–82, 86–90, 92–95, 98–104, 107–108, 110–111, 113–114, 116, 120–121, 124, 130, 132; logic of 5–6, 12, 26, 45, 77, 94, 108
Cooper, Melinda 25
Crary, Jonathan 41
Csicsery-Ronay, Istvan 11
cyberpunk 2–4, 6, 9–13, 15–20, 23–36, 38, 42–45, 82, 86, 88, 90, 93, 100–104, 107, 129–131
cyborg 11, 16–19, 23–24, 28, 30–31, 33, 36, 42–45, 49, 58–60, 65, 80, 89, 104, 107–108, 110, 118, 129, 131–132
cyborganization 17, 19, 23–24, 28–29, 31, 45, 84, 130

Index 135

Davis, Mike 10, 27, 103
death 2–11, 13, 19, 26–30, 35, 43, 48, 65–70, 72–82; as death-making 79; as death-worlds 6–8, 67, 78, 93; ordinary 66, 79; slow 77–78, 83; social 79–80
defamiliarize 9, 16, 18, 19, 24, 129–130
Deleuze, Gilles 4, 31–32, 87–90, 112–113
Deleuzian Thatcherism 111, 113

economization 2–3, 7, 28, 42, 90, 111, 116, 118
entrepreneur of self 2, 31, 41–42, 47, 50, 52, 55, 59, 109

familiarize 9, 24
Fanon, Frantz 73
Fitbit 45, 50–59
flexible work 5, 35, 80, 131
Foucault, Michel 3–4, 6–7, 9, 13, 17–19, 25–26, 29, 45, 47–50, 52, 54–55, 58–60, 65–77, 79–81
freedom 27–28, 35–36, 92, 100–101

genocide 4, 65, 71, 76
governmentality 1–2, 4–5, 8–9, 11–13, 42, 45, 49–52, 55, 65, 75, 78, 81, 112, 116 ; biopolitical 75, 78, 109; as governmental rationality 2, 7, 13, 19, 76, 101, 130–132; neoliberal 1–2, 5, 8–9, 12–13, 42, 45, 49–52, 55, 65, 81, 112, 116
Graham, Stephen 6
Guattari, Félix 112–113, 125

Haraway, Donna 49, 87
Harvey, David 3, 11–13, 15–16, 88, 91–92, 130
health 1, 6–9, 12, 17, 29, 31, 42–55, 69–70, 72–73, 78, 80–81, 89–90, 93, 109–110, 119, 123, 131; neoliberal 17, 42–43, 89, 119
Holmes, Ryan 47
Huawaei 44
human capital 1–2, 5–6, 9, 11, 17, 19, 26–28, 31, 36, 41–44, 47, 49, 54, 57–59, 65, 76, 80, 82, 87–90, 98, 101–103, 108–110, 114, 119, 131–132
hyperhuman 29–30
hyperobject 34, 36, 88–89, 130

illegal 9–10, 15, 31, 88, 97–98, 101–103

individual 1–3, 5–19, 24–29, 31, 35–36, 38, 41–52, 54–58, 65, 69, 71–72, 74, 76–82, 87–94, 96, 98–104, 108–111, 113–117, 119–120, 130–132; possessive 1–2, 8–9, 35, 101, 108, 110, 114
insecurity 1–3, 8, 12–13, 25, 27–32, 35, 72, 75–81
instrumental rationality 12–13, 36, 130
intense 5, 7, 18–19, 43–44, 46, 55, 57, 65, 72, 92, 96–97, 99, 110, 120, 122, 125, 131; intensely 2, 5, 8, 16–19, 42, 44–45, 58, 60, 87, 92–93, 97, 99, 107–108, 110, 112, 116–117, 119–122, 124–125, 129, 131–132; intensified 5–6, 18–19, 33, 38, 43, 72, 80, 94, 100, 104, 109, 114–116, 119, 122, 131; intensity 2, 16–19, 26, 41, 44, 48, 89–93, 95–96, 99, 104, 107–108, 110, 120, 122, 129, 132; as intensification 6, 17–19, 33, 38, 41, 43, 45, 89, 92, 107–111, 113–114, 116, 118, 124–125

James, Robin 44
Jameson, Fredric 3, 11–16, 24, 34, 86–88, 90–92, 95, 97, 99–100, 103, 130

Kelly, Kevin 46
killing 1–2, 4, 7–8, 13, 18, 29, 65–81, 93, 94, 98, 99, 132–133; massive-scale 4, 18, 67; spectacular 1, 4, 7, 65–68, 70, 72, 76–77, 79, 81, 94, 131, 133; sub-state 7, 66, 71, 79; subtle state 1, 8, 12–13, 16, 29, 66–67, 72, 77, 93–94

Lacan, Jacques 95
Land, Nick 107, 111–114
Lemke, Thomas 1, 17, 27, 42, 52
liberalism 14, 24–25, 42, 75, 79, 113, 131
Luke, Tim 14, 23–24, 33

Mahoney, Elizabeth 100
marketization 6, 10, 23, 65, 89, 93, 112, 114
Marxist 11–12, 41
Mbembe, Achille 6–8, 65–78, 80–81, 93–94, 97
McCaffery, Larry 12
militarization 6, 10

Index

narrative(s) 2–3, 9, 14–15, 18, 23–25, 31–32, 34, 45, 50, 89, 95–97, 100, 103, 129–131
Nealon, Jeffrey 44
necropolitics 4, 6, 18–19, 65–67, 72, 75–76, 78, 93, 132–133; necropower 6, 66–67, 73–74, 78, 80, 93
necroscapes 5–6, 8, 93–94, 101, 103, 129, 132
necro-temporality 5, 10, 82, 90–91, 93–94, 96–99, 101, 103, 129, 132
neoliberalism 1–5, 8–11, 13–14, 16–19, 23–31, 34, 36, 41–45, 48–49, 52, 54, 57–60, 65–67, 72, 75–82, 86–104, 107–110, 112–118, 120–122, 124–125, 129–132
neoliberal subject 5, 12, 17, 19, 27, 36, 43, 45, 52–55, 58–60, 81–82, 89, 104, 107–108, 116, 119–120, 124
Neuromancer 26–27, 32–35, 43
normalize 3, 8–9, 13, 43, 45, 47, 50–52, 56–57, 66, 70, 80, 90, 123
Noys, Benjamin 28, 34, 111–114, 116

pastiche 11, 90–91, 99–100
population/s 1, 3, 5–13, 16–17, 19, 29–31, 33, 42–45, 48–50, 65, 67, 69–70, 72–73, 75–82, 86–90, 92–95, 98, 101–103, 114, 120, 131–132
postmodern 3–4, 9, 11–16, 86–88, 90–92, 95–97, 99–100, 103, 130; postmodernism 11–12, 14–15, 91–92, 95
Povinelli, Elizabeth 4, 75–77, 82, 131
Precarity 2–3, 5, 10, 12; precarious 5, 8, 10, 14, 17, 80, 94, 101, 122, 125, 131
project cyborg 118

racism 1–2, 4, 9, 12, 29, 31, 35, 65, 67–80, 95, 98, 101, 131; biological 4, 29, 35, 65, 69, 131; systemic 2, 12
replicants 15, 30–31, 91–92, 95–99, 101–103, 110
resilience 2, 8, 16–19, 46, 48, 76, 92, 99–102, 104, 107–108, 110–111, 116–119, 121, 125, 129, 131–132; resilient 16, 18–19, 92, 99, 101–104, 108, 110, 116–118, 120–121, 124, 129, 131

responsibility 2, 5–6, 9–10, 13, 17–19, 25–27, 29, 31, 42–44, 48–49, 52, 54, 60, 65, 78–81, 90, 92, 94, 119, 130, 132
risk 2, 4, 6, 10–11, 13, 19–20, 25, 35, 43–44, 48, 60, 65, 72, 76, 78–79, 81, 86, 89–90, 92–94, 107–110, 113–114, 122, 131

Schmitt, Carl 8, 94
science fiction 2–3, 9, 11, 16–17, 23–24, 34, 36, 42, 84, 87, 104, 117–118, 124, 129–130
Scott, Ridley 91–92, 102–103
security 5, 7, 10, 23, 27–28, 32, 73, 75, 78, 98, 101–103, 131–132
self-cultivation 17, 19, 36, 41–45, 53, 57, 59–60, 108–109, 117–120, 124, 129, 131; and maximization 6, 41, 45–46, 55, 58, 80, 118, 120; and monitoring 17–19, 36, 41–49, 51–52, 55–56, 58–60, 65, 80, 89, 107–110, 119, 129, 131–132
Shapiro, Michael 3, 23
Shaviro, Steven 34, 36, 43, 59, 86–90, 111–112, 114, 116–118, 124, 129
simulacrum 11
sleep 41–42, 44, 47, 51, 56, 89, 98, 119
Snow Crash 27, 33, 35
Social Darwinism 27, 92
Srnicek, Nick 107–108, 114–116, 121–124
surplus bodies 1, 5, 10, 12–14, 29–30, 79, 97, 117, 124, 129
Suvin, Darko 86

thanatopolitics 4, 67, 75
Traywick, Aaron 109, 122–123

Urban 1, 4, 6–10, 16–17, 35, 73, 78, 80–81, 91, 93–94, 101–102, 108

Wacquant, Loïc 25–26, 32, 76, 80–81
Warwick, Kevin 118
Williams, Alex 107–108, 114–116, 121–124
Wolf, Gary 46–47, 56

Zayner, Josiah 121–123